AF790531

Livestock Health and Farming

Edited by Muhammad Abubakar

Published in London, United Kingdom

IntechOpen

Supporting open minds since 2005

Livestock Health and Farming
http://dx.doi.org/10.5772/intechopen.77836
Edited by Muhammad Abubakar

Contributors
Albert Sundrum, Cristina Villamar, Cristóbal Sardá, Hr Meena, Vikash Kumar, Letha Devi, Anjumoni Mech, Veerasamy Sejian, Ravikiran G, Mukund Kataktalware, Clarice Mudzengi, Everson Dahwa, Clayton Kapembeza, Gerardo A. Pámanes-Carrasco, Esperanza Herrera-Torres, Manuel Murillo-Ortiz, Damián Reyes-Jáquez, Okanlade Lawal-Adebowale, Muhammad Abubakar

© The Editor(s) and the Author(s) 2020
The rights of the editor(s) and the author(s) have been asserted in accordance with the Copyright, Designs and Patents Act 1988. All rights to the book as a whole are reserved by INTECHOPEN LIMITED. The book as a whole (compilation) cannot be reproduced, distributed or used for commercial or non-commercial purposes without INTECHOPEN LIMITED's written permission. Enquiries concerning the use of the book should be directed to INTECHOPEN LIMITED rights and permissions department (permissions@intechopen.com).
Violations are liable to prosecution under the governing Copyright Law.

Individual chapters of this publication are distributed under the terms of the Creative Commons Attribution 3.0 Unported License which permits commercial use, distribution and reproduction of the individual chapters, provided the original author(s) and source publication are appropriately acknowledged. If so indicated, certain images may not be included under the Creative Commons license. In such cases users will need to obtain permission from the license holder to reproduce the material. More details and guidelines concerning content reuse and adaptation can be found at http://www.intechopen.com/copyright-policy.html.

Notice
Statements and opinions expressed in the chapters are these of the individual contributors and not necessarily those of the editors or publisher. No responsibility is accepted for the accuracy of information contained in the published chapters. The publisher assumes no responsibility for any damage or injury to persons or property arising out of the use of any materials, instructions, methods or ideas contained in the book.

First published in London, United Kingdom, 2020 by IntechOpen
IntechOpen is the global imprint of INTECHOPEN LIMITED, registered in England and Wales, registration number: 11086078, 7th floor, 10 Lower Thames Street, London, EC3R 6AF, United Kingdom
Printed in Croatia

British Library Cataloguing-in-Publication Data
A catalogue record for this book is available from the British Library

Additional hard and PDF copies can be obtained from orders@intechopen.com

Livestock Health and Farming
Edited by Muhammad Abubakar
p. cm.
Print ISBN 978-1-78985-903-4
Online ISBN 978-1-78985-904-1
eBook (PDF) ISBN 978-1-78984-636-2

For EU product safety concerns:
IN TECH d.o.o., Prolaz Marije Krucifikse Kozulić 3, 51000 Rijeka, Croatia, info@intechopen.com or visit our website at intechopen.com.

We are IntechOpen, the world's leading publisher of Open Access books

Built by scientists, for scientists

4,800+
Open access books available

122,000+
International authors and editors

135M+
Downloads

151
Countries delivered to

Our authors are among the
Top 1%
most cited scientists

12.2%
Contributors from top 500 universities

WEB OF SCIENCE™

Selection of our books indexed in the Book Citation Index
in Web of Science™ Core Collection (BKCI)

Interested in publishing with us?
Contact book.department@intechopen.com

Numbers displayed above are based on latest data collected.
For more information visit www.intechopen.com

Meet the editor

Dr. Muhammad Abubakar is a senior scientist with the National Veterinary Laboratory, Islamabad, Pakistan, who has worked for more than 15 years in veterinary sciences. He has expertise in transboundary animal diseases (TADs) and has worked with local and international projects for the establishment of diagnostic laboratories for TADs. He has also conducted numerous trainings for field as well as laboratory staff. Dr. Abubakar has supervised various research studies and published numerous research papers, review articles and book chapters. He is co-editor of *The Role of Biotechnology in Improvement of Livestock* and editor-in-chief for two veterinary science journals.

Contents

Preface

It is important to recognise recent advances in animal health and related sciences to identify and control diseases and disorders in animals. *Livestock Health and Farming* provides a detailed description of key aspects of livestock health issues and farming practices. It covers introductory and intermediate topics for teachers, veterinary students, and practitioners in the field. Chapters are written by renowned experts and organized into the following sections:

- Livestock Health
- Livestock Farming and Nutrition
- Livestock Farming and Climate Change

The first section covers farming and the overall health of livestock. There is a particular focus on antimicrobial resistance in livestock, which is especially important for developing countries where there are no strict rules restricting the use of antibiotics.

In the second section, there are three chapters covering nutrition and its role in animal health and farming. Chapters discuss nutrition and health management in dairy animals, the nutrient cycle during livestock production, and livestock feed and feeding in semi-arid regions.

The third section includes chapters on the unique aspects of nutrient, health, and other requirements of animal farming.

I would like to express my gratitude to all the contributors of this book including the authors of the accepted chapters. My special thanks to Author Service Manager Manager, Ms. Jasna Bozic, and other staff of IntechOpen for their kind support and great efforts in bringing the book to completion. I am also thankful to my colleagues, friends, and family for all their support in the fulfilment of this project.

Dr. Muhammad Abubakar
National Veterinary Laboratory,
Islamabad, Pakistan

Section 1

Livestock Health

Chapter 1

Introductory Chapter: Livestock Health and Farming - Regional to Global Perspectives

Muhammad Abubakar, Abdullah Iqbal, Shumaila Manzoor and Muhammad Javed Arshed

1. Introduction

Livestock comprises a global asset of more than $1.4 trillion. The livestock sector is important in both developed and developing counties. Almost 1.3 billion peoples are involved with the livestock sector directly or indirectly. Animals are an important source of nutrients in the form of meat and milk. Livestock products provide 33% of total protein intake throughout the world [1].

Livestock is one of the major subsectors of agriculture that is growing rapidly because of the increase in demand for livestock products [2]. According to estimates, global meat production would increase to 465 million tons by 2050 from 229 million tons in 1999. Similarly, milk production is expected to increase to 1043 million tons by 2050 from 580 million tons in 1999 [3].

There is a vast difference between the livestock sector of developed countries and developing countries [4]. There are more chances of an increase in the value of livestock in developing countries due to growing demand, but in industrialized countries, demand is stagnant. To meet this requirement, farmers should work on the vertical expansion of livestock [5].

The livestock sector has many environmental impacts. Globally, the livestock sector is the second-largest pollution-producing sector after the electricity industry [6]. Animals are responsible for emissions of gasses such as ammonia (NH_3), carbon dioxide (CO_2), methane (CH_4), and nitrous oxide (N_2O). These gasses are the cause of global warming and acid rains [7].

The health of animals can be described as normal physiological functioning of all the systems of the body of animals to achieve the highest production or the lack of disease. Whenever an animal gets ill, economic issue arises. Livestock diseases result in loss of production, treatment cost, prevention cost, and a barrier to trade [8].

In developed countries, any livestock disease outbreak would affect the economy of the farm and country. On the other hand, in developing countries in case of any livestock disease outbreak, additional factors like food scarcity, loss of draught power, and social security are also emphasized [9]. Public health is also an issue related to livestock disease as many of these diseases are zoonotic. Transmission of these diseases takes place either by direct contact (tuberculosis, brucellosis) or via vectors (Lyme disease, West Nile disease, Rift Valley fever) [10]. Spread of antimicrobial resistance because of the misuse and abuse of antibiotics in the livestock sector is also an iceberg that we are facing today [11].

2. Animal production systems

Animals are reared under different types of production systems throughout the world. Animal production systems are mostly categorized on the basis of capital investment and outputs. The first and most primitive animal production system is an exploitation production system, which is categorized by no capital investment and minimal human labor. Animals depend upon environmental resources and outputs are highly variable [12].

Animal production system is an extensive production system. This production system is categorized by minimal inputs and outputs. The survival of animals is important as compared to peak productivity. This system is a not market-oriented system. Subsistence farming and ranching are two examples of the extensive production system. In subsistence farming, animals are kept to fulfill the need of the family of a farmer [13]. Sometimes extra products are sold. Ranching includes large herds of animals grazing here and there. Treatment or vaccinations are performed rarely in the extensive production system [14].

The intensive animal production system is featured by high input and high output. Animals are kept in a favorable environment and all the nutrition requirements are met. Animals are vaccinated according to schedule. Prompt treatment of diseased animals is also a silent feature of intensive farming. To get maximum production from animals is the main aim of this system [14].

3. Effect of climate change on livestock health

Climate change has the potential to affect animal health both positively and negatively. Climate change can affect animal health directly, indirectly, and by altering environmental conditions. The direct impact of climate change on animal health is manifested by an increase in temperature and heat waves [15]. Heat stress in animals can cause metabolic problems, immunosuppression, and oxidative stress. These complications can eventually lead to the death of the affected animal [16].

Indirect impacts due to climate change on animal health can be due to change of microbial distribution or density, the incidence of vector-borne diseases, and water and food scarcity. For instance, a slight temperature change can alter the relative humidity and promote the reproduction of insects. These insects can act as a vector of many protozoan and viral diseases [17].

Climate change also affects the parasitic diseases of animals as most of the gastrointestinal parasites of animals live only a short life span inside the body of the animal and most of the remaining life cycle of these parasites is completed outside the body of their host. So, the life cycle of gastrointestinal parasites is affected by climate change [15, 18].

4. Impact of livestock movement and trade on disease incidence

Infectious diseases can be transmitted from one host to other susceptible hosts in different ways like direct contact, vector-borne, and airborne transmission. Animal diseases can be controlled by decreasing the direct contact of healthy animals with sick ones. Trade of livestock and its products is a complex process [19].

Although the OIE has set standards about animal health and the spread of zoonotic diseases, still, most of the livestock trade is based on the bilateral agreement of countries. Two main areas of focus by the OIE in livestock trade are animal health and the welfare of animals. Movement of animals can introduce exotic animal diseases or zoonotic pathogens [20, 21].

FMD outbreak of 2001 in England is one of the major examples of the spread of disease by animal movement. The first outbreak of FMD was reported in February 2001 from the North of England. Within 2 months, FMD virus had spread to France and the Netherlands by the transportation of animals from England. This outbreak resulted in the loss of almost £8 billion to the public and private sectors [1, 22].

Another example of the spread of disease by the movement of animals is the rabies outbreak in Flores Island in Indonesia. Until 1997, Flores Island was rabies-free. Rabies outbreak was reported after the import of three dogs from rabies-endemic area. This outbreak resulted in the death of 113 humans and 50% of the dog population was culled. But still, rabies is endemic in Flores Island [23, 24].

Markets play a very important role in the spread of livestock diseases. It is vital to understand the role of quarantine measures and risks associated with the movement of animals. Spread of diseases by animal movements and trade is not the issue of any one country, rather it is a global problem. Following standards for animal trade and global cooperation can help in minimizing the disease spread by animal movements [1, 20].

5. Political instability and livestock health

Political stability and food security of any country are directly linked. A politically stable country would be superior to a country that is politically unstable in terms of food availability, the health of humans and animals, and education [1]. In any emergency situation, migration of people and animals would increase the chances of zoonotic outbreaks. For instance, during the Gulf War, rinderpest was introduced in Turkey by the migration of animals from Iraq. Rinderpest outbreak caused panic in Turkish farmers too, and they started to sell their sick animals in the markets. This panic approach further spread the disease [25].

Before 2014 uprising Syria was at stage 3 of Health "Progressive Control Pathway" for FMD. But after that due to lack of veterinary services, that status is lost. Vaccination of animal herds against brucellosis was also impeded by this revolt which resulted in a marked increase in brucellosis incidence in the human population [26]. In 2014, political instability in Nigeria became one of the reasons for the re-emergence of H5N1 in poultry. Because of improper control strategies, H5N1 outbreaks were also documented in the next 2 years from neighboring countries [1].

6. The social effects of livestock diseases

The diseases of livestock can affect the community at different levels varying from an individual farmer to a multinational food chain. Due to livestock disease in any area, society would be affected either directly or indirectly. A direct effect of livestock disease includes the zoonotic aspect and possible morbidity and mortality due to these zoonotic pathogens. An indirect effect of livestock diseases includes financial burden and social and indirect health consequences [27].

In developing countries, livestock farming is not only their business but the way of their life. Livestock diseases in developing countries where most of the livestock is kept under subsistence farming can seriously affect the social norms [28]. The decrease in livestock production would affect the total income of farmers and hence living standards. Indirect health consequences due to animal disease include stress, anxiety, and depression to livestock producers [29]. For example, culling of animals due to the outbreak of bovine spongiform encephalitis in Canada resulted in stress and guilt feeling in owners of those animals. Many of these livestock owners were keeping inherited herd [27].

7. Role of biosecurity in livestock farming

Biosecurity at the farm level involves all the steps taken to limit the entry of pathogens and the spread of disease at the farm. Biosecurity applies to both contact of farm animals with other live animal and indirect contact with any contaminated vehicle or equipment [30]. External biosecurity measures are those that are used to decrease the chances of entry of pathogens into a farm. Internal biosecurity includes steps taken to prevent the spread of disease to healthy animals within a farm from diseased animals [31].

Enforcement of biosecurity at every farm and country border is a global responsibility for combating threats like food security and animal and human diseases. Good farm biosecurity can play a vital role in minimizing the outbreaks of both endemic and exotic diseases [32].

Farm biosecurity is based on four principles. The first is limiting the introduction of new animals in the herd and adopting quarantine measures. The second biosecurity principle is controlled movements of people and vehicle and equipment hygiene. The provision of feed and water that is free from pathogenic contamination is the third principle. The fourth principle is the regular vaccination of animals against endemic diseases along with accurate surveillance and reporting of transboundary animal diseases [33].

8. Conclusion

Animal health is directly linked to food security. Now the world should accept the fact that animal diseases are not a problem of any specific country or region, rather they are global issues. Developing countries should adopt the international standard for the trade of animals and augment the disease surveillance system. For a better future of the world, decision-makers should turn their attention toward the food-producing system and epizootics.

Author details

Muhammad Abubakar*, Abdullah Iqbal, Shumaila Manzoor and Muhammad Javed Arshed
National Veterinary Laboratory, Pakistan

*Address all correspondence to: hayee42@yahoo.com

IntechOpen

© 2020 The Author(s). Licensee IntechOpen. This chapter is distributed under the terms of the Creative Commons Attribution License (http://creativecommons.org/licenses/by/3.0), which permits unrestricted use, distribution, and reproduction in any medium, provided the original work is properly cited.

References

[1] Lubroth J, Idrissi AE, Myers L, Hasibra M, Black P, Burgeon D. Linking animal diseases and social instability. Revue Scientifique Et Technique De LOIE. 2017;**36**(2):445-457. DOI: 10.20506/rst.36.2.2665

[2] Thornton PK. Livestock production: Recent trends, future prospects. Philosophical Transactions of the Royal Society, B: Biological Sciences. 2010;**365**(1554):2853-2867. DOI: 10.1098/rstb.2010.0134

[3] Dopelt K, Radon P, Davidovitch N. Environmental effects of the livestock industry: The relationship between knowledge, attitudes, and behavior among students in Israel. International Journal of Environmental Research and Public Health. 2019;**16**(8):1359. DOI: 10.3390/ijerph16081359

[4] Morand S. Impact of climate change on livestock disease occurrences. In: Climate Change Impact on Livestock: Adaptation and Mitigation. New Delhi: Springer; 2015:113-122. DOI: 10.1007/978-81-322-2265-1_8

[5] Reynolds LP, Wulster-Radcliffe MC, Aaron DK, Davis TA. Importance of animals in agricultural sustainability and food security. The Journal of Nutrition. 2015;**145**(7):1377-1379. DOI: 10.3945/jn.115.212217

[6] Grossi G, Goglio P, Vitali A, Williams AG. Livestock and climate change: Impact of livestock on climate and mitigation strategies. Animal Frontiers. 2018;**9**(1):69-76. DOI: 10.1093/af/vfy034

[7] Rojas-Downing MM, Nejadhashemi AP, Harrigan T, Woznicki SA. Climate change and livestock: Impacts, adaptation, and mitigation. Climate Risk Management. 2017;**16**:145-163. DOI: 10.1016/j.crm.2017.02.001

[8] Ducrot C, Bedhom B, Béringue V, Coulon J-B, Fourichon C, Guérin J-L, et al. Issues and special features of animal health research. Veterinary Research. 2011;**42**(1):96. DOI: 10.1186/1297-9716-42-96

[9] Rich KM, Perry BD. The economic and poverty impacts of animal diseases in developing countries: New roles, new demands for economics and epidemiology. Preventive Veterinary Medicine. 2011;**101**(3-4):133-147. DOI: 10.1016/j.prevetmed.2010.08.002

[10] Tomley FM, Shirley MW. Livestock infectious diseases and zoonoses. Philosophical Transactions of the Royal Society, B: Biological Sciences. 2009;**364**(1530):2637-2642. DOI: 10.1098/rstb.2009.0133

[11] Ayukekbong JA, Ntemgwa M, Atabe AN. The threat of antimicrobial resistance in developing countries: Causes and control strategies. Antimicrobial Resistance and Infection Control. 2017;**6**(1). DOI: 10.1186/s13756-017-0208-x

[12] Pingali PL. From subsistence to commercial production systems: The transformation of Asian agriculture. American Journal of Agricultural Economics. 1997;**79**(2):628-634. DOI: 10.2307/1244162

[13] Temple D, Manteca X, Velarde A, Dalmau A. Assessment of animal welfare through behavioural parameters in Iberian pigs in intensive and extensive conditions. Applied Animal Behaviour Science. 2011;**131**(1-2):29-39. DOI: 10.1016/j.applanim.2011.01.013

[14] Castel JM, Mena Y, Delgado-Pertíñez M, Camúñez J, Basulto J, Caravaca F, et al. Characterization of semi-extensive goat production Systems in Southern Spain. Small Ruminant

Research. 2003;**47**(2):133-143. DOI: 10.1016/s0921-4488(02)00250-x

[15] Lacetera N. Impact of climate change on animal health and welfare. Animal Frontiers. 2018;**9**(1):26-31. DOI: 10.1093/af/vfy030

[16] Akbarian A, Michiels J, Degroote J, Majdeddin M, Golian A, Smet SD. Association between heat stress and oxidative stress in poultry; mitochondrial dysfunction and dietary interventions with phytochemicals. Journal of Animal Science and Biotechnology. 2016;**7**(1). DOI: 10.1186/s40104-016-0097-5

[17] Bernabucci U. Climate change: Impact on livestock and how can we adapt. Animal Frontiers. 2019;**9**(1):3-5. DOI: 10.1093/af/vfy039

[18] Rust JM. The impact of climate change on extensive and intensive livestock production systems. Animal Frontiers. 2018;**9**(1):20-25. DOI: 10.1093/af/vfy028

[19] Bate AM, Jones G, Kleczkowski A, Naylor R, Timmis J, White PCL, et al. Livestock disease management for trading across different regulatory regimes. EcoHealth. 2018;**15**(2):302-316. DOI: 10.1007/s10393-018-1312-y

[20] Perry BD, Grace D, Sones K. Current drivers and future directions of global livestock disease dynamics. Proceedings of the National Academy of Sciences. 2011;**110**(52):20871-20877. DOI: 10.1073/pnas.1012953108

[21] Brooks-Pollock E, Jong MD, Keeling M, Klinkenberg D, Wood J. Eight challenges in modelling infectious livestock diseases. Epidemics. 2015;**10**: 1-5. DOI: 10.1016/j.epidem.2014.08.005

[22] Blake A, Sinclair MT, Sugiyarto G. Quantifying the impact of foot and mouth disease on tourism and the UK economy. Tourism Economics. 2003;**9**(4):449-465. DOI: 10.5367/000000003322663221

[23] Wera E, Mourits MCM, Hogeveen H. Uptake of rabies control measures by dog owners in Flores Island, Indonesia. PLoS Neglected Tropical Diseases. 2015;**9**(3). DOI: 10.1371/journal.pntd.0003589

[24] Putra AAG, Hampson K, Girardi J, Hiby E, Knobel D, Mardiana W, et al. Response to a Rabies Epidemic, Bali, Indonesia, 2008-2011. Emerging Infectious Diseases. 2013;**19**(4):648-651. DOI: 10.3201/eid1904.120380

[25] Food and Agriculture Organization of the United Nations (FAO). The impact of political and social instability on animal health. FAO Animal Production and Health Papers. 2002;**153**:47

[26] Food and Agriculture Organization of the United Nations (FAO). FAO/OIE progressive control pathway for foot and mouth disease. 2011. Available from: www.fao.org/fileadmin/user_upload/eufmd/docs/PCP/PCP_Guidelines_Eng_2012web.pdf [Accessed: 08 January 2020]

[27] Evans B. The social and political impact of animal diseases. Veterinaria Italiana. 2006;**42**(4):399-406

[28] Kaplan M. Social effects of animal diseases in developing countries. Bulletin of the Atomic Scientists. 1966;**22**(9):15-21. DOI: 10.1080/00963402.1966.11454985

[29] Crimes D, Enticott G. Assessing the social and psychological impacts of endemic animal disease amongst farmers. Frontiers in Veterinary Science. 2019;**6**. DOI: 10.3389/fvets.2019.00342

[30] Robertson ID. Disease control, prevention and on-farm biosecurity: The role of veterinary epidemiology.

Engineering. 2020;**6**(1):20-25. DOI: 10.1016/j.eng.2019.10.004

[31] Brennan M, Kemp R, Christley R. Direct and indirect contacts between cattle farms in north-West England. Preventive Veterinary Medicine. 2008;**84**(3-4):242-260. DOI: 10.1016/j.prevetmed.2007.12.009

[32] Nöremark M, Sternberg-Lewerin S. On-farm biosecurity as perceived by professionals visiting Swedish farms. Acta Veterinaria Scandinavica. 2014;**56**(1). DOI: 10.1186/1751-0147-56-28

[33] Higgins V, Bryant M, Hernández-Jover M, Rast L, Mcshane C. Devolved responsibility and on-farm biosecurity: Practices of biosecure farming care in livestock production. Sociologia Ruralis; 2016;**58**(1):20-39. DOI: 10.1111/soru.12155

Chapter 2

Antimicrobial Resistance and Rational Use of Antimicrobials in Livestock: Developing Countries' Perspective

Hans Ram Meena and Vikash Kumar

Abstract

India is one of the top consumers of antibiotics in agriculture worldwide, which accounts for 3% of global consumption, which is estimated to double in 2030. The use of antibiotics, particularly in chickens, is expected to triple in India by 2030. The overuse, injudicious use, and misuse of these antimicrobial drugs have spawned the evolution of life-threatening bacteria that is making the current antimicrobials' reserve useless. Suitable extension outreach and continuing programmes should be devised to promote the judicious use of antimicrobials. Innovative approaches, such as One Health, Antimicrobial Stewardship, and antimicrobial conservation are the need of present alarming situation. There is need to reduce the antimicrobial use in animals, particularly domesticated animals; provision of infection surveillance in hospitals; improving hospital surveillance for monitoring antibiotic resistance; promoting rational and judicious use of drug through education, monitoring, and supervision; researching new drugs; and developing and implementing a more restrictive and participatory antibiotic policy by including various stakeholders. Thus, tracking the rate of veterinary antimicrobial use, resistance, and residues, through a nationwide surveillance and monitoring system, and educating farmers, veterinarians, and consumers could pave the way to fight against this catastrophic situation of antimicrobial resistance.

Keywords: antimicrobial resistance, antimicrobial stewardship, One Health, stakeholders, surveillance

1. Introduction

Antimicrobial agents are widely used in food-animal production for disease prevention and treatment in animals, to control disease spread, to prevent contamination of the food chain via horizontal and vertical transfer of antimicrobial resistance, and to increase productivity [1]. However, their overuse in humans and animals leads to the emergence of antimicrobial resistance, a general term that encompasses decreased and poor efficacy of antimicrobials to treat disease [2]. Recent projections revealed that by 2050 global livestock production would fall by 3–8% each year, as result of which annual global gross domestic product will decline by 1.1–3.8%. Due to rise in disease incidence, low income countries will be affected more severely, with

IntechOpen

a predicted rise of extremely poor people from 6.2 to 18.7 million by 2030 [3]. Rise in frequency of treatment failures have been reported in treatments with infections caused by multi-, extensive-, and pan-drug resistant bacteria. Once antimicrobials (antibiotics) normally used against bacteria lose their efficacy to treat disease, it becomes necessary to use others, so-called "reserve" or "last resort" options that are often more expensive and/or toxic preparations [4]. In several developing countries, antimicrobial consumption is expected to rise considerably due to increase in meat consumption, from Indonesia (202%) and Nigeria (163%) to Vietnam (157%) and Peru (160%), by 2030 [5]. Organization for Economic Cooperation and Development (OECD) estimated that antimicrobials used in food-animal production will increase by 67% globally, i.e., from 63,000 in 2010 to 106,000 tonnes by 2030—an increase of 67% [6]. Thus, overuse of antimicrobials in the food-animal production sector gives rise to antimicrobial resistance in animal pathogens, leading to increase in therapy failure with a negative effect on animal health and welfare [7]. The immediate cost of withdrawal of non-therapeutic antimicrobials at animal level, without adjustments in production processes, may decrease the feed efficiency, growth, survival, and number of animals born [8].

2. Genesis of antibiotic resistance and its spread across geographical boundaries

The World Health Organization (WHO) has emphasized the need for an integrated and coordinated global effort to control antibiotic resistance. In 2001, the World Health Organization Global Strategy for Containment of Antimicrobial Resistance has provided a framework of interventions to slow the emergence and reduce the spread of antimicrobial-resistant microorganisms across geographical boundaries and species [9]. For understanding the genesis and spread of antimicrobial resistance across species and increase in resistosome burden, the following sub-heads points the focus.

2.1 Development and spread of antimicrobial resistance

The development of resistance in microbes arises in two ways: (i) intrinsic resistance, which occurs when the microbial species is able to innately resist the activity of an antimicrobial agent (by preventing either the entry or binding of the antimicrobial agent); and (ii) acquired resistance, in which once-susceptible microbial species mutate or obtain genes from other microbe, to acquire resistance. Antimicrobial resistance cannot be prevented because every time antimicrobials are used, the effective lifespan of that antimicrobial drug is shortened [10]. In general, few categories of pathogen are responsible for a large portion of resistant infections in humans. One of them is New Delhi metallo-β-lactamase-1 (NDM-1) gene which confers broad resistance to most antibiotics, including carbapenems, and can be transferred to a wide variety of bacterial species [11]. Another is resistant Gram-negative bacteria which carry extended-spectrum beta-lactamase enzymes (ESBLs), responsible for high levels of resistance to some of the most commonly prescribed antibiotics [12].

2.2 Antibiotic use in livestock and resistance

Livestock contributes for over a fourth of India's total agricultural output, and 4% of the gross domestic product (GDP). India is one of the top consumers of antibiotics in agriculture worldwide, which accounts for 3% of global consumption, which is estimated to double in 2030 [13]. Resistant microbes and residues have been detected in living bovines, chickens, honey, pigs, horses, donkeys and mules, and fish and shellfish. In cattle, resistant strains of coagulase-negative staphylococci, *Escherichia coli*, and *Staphylococcus aureus*, extended-spectrum beta-lactamase (ESBL) and New Delhi metallo-beta-lactamase (NDM-1) genes, have been reported. *E. coli*, *S. aureus*, enterococci, *Pasteurella multocida*, *Campylobacter jejuni*, and *Salmonella*, including ESBL-producing strains have been found in poultry. The chances for antimicrobial resistant microbes in the race for survival are in direct proportion to the volume of antibiotics used, this makes it more critical to examine current habits and encourage rational and conservative use of antimicrobials. Due to antimicrobial resistance, easy-to-treat infections are becoming difficult or impossible to cure, with an unambiguous global increase in both livestock mortality and treatment costs [12].

2.3 Use of antimicrobials for different purposes

Therapeutic use of antimicrobials is meant for treatment of diseases. However, if a few animals are found to be sick, often the whole flock or herd will be treated (known as meta-phylaxis or sub-therapeutic) to prevent the disease spreading. Thus, there is not always a clear distinction between treatment and prevention [14]. In this condition, treatment usually occurs at high doses for a relatively short period of time. Prophylactic treatment is done for prevention of disease. The treatment of animals is done with low, sub-therapeutic doses of antibiotics via feed or drinking water, even in the absence of any signs of disease but when there is risk of infection. Treatment can be given over a period of several weeks, and sometimes longer. Antibiotics are also used for growth promotion. Here, very low sub-therapeutic doses of antibiotics are given to animals (particularly intensively kept pigs and poultry) in their feed, in order to increase their growth-rate and productivity. Treatment is continuous and it lasts for a long time [15].

2.4 Use of antimicrobials in food animals

A study revealed that annually, 45, 148, and 172 mg/kg antimicrobials are consumed by cattle, chicken, and pigs, respectively, to produce each kilogram of their meat. The global consumption of antimicrobials estimated to increase by 67% from 2010 to 2030, i.e., from 63,151 ± 1560 to 105,596 ± 3605 tons [16]. At present time, more antibiotics are used worldwide in poultry, swine, and cattle production than in the entire human population [12]. In aquaculture, antibiotics are used for therapeutic and prophylactic purposes often in high concentrations because bacteria travel in water easily, here antibiotics are not used for growth promotion. In the BRICS countries (Brazil, Russia, India, China and South Africa), antibiotic use in animals is expected to double by 2030. Use of antibiotics, particularly in chickens, is expected to triple in India by 2030 [16].

2.5 Use of antibiotics in dairying

The antibiotic residues are at alarming rate in dairying in India. A study by Ramakrishna and Singh [17] in 1985 revealed that streptomycin was found in 6% milk samples in Haryana. One decade later, in Hyderabad, Secunderabad, and surrounding villages dairy farmers were surveyed on antibiotic use practices. Among 38 dairy farmers, about 50% of them used oxytetracycline to treat diseases such as mastitis and fever; the survey revealed that oxytetracycline residues were found in 9% samples from markets and 73% individual animals, while no residues were found in government dairy samples [18]. A survey conducted by the National Dairy

Research Institute near Bangalore in 2000 revealed that tetracyclines, gentamycin, ampicillin, amoxicillin, cloxacillin, and penicillin were commonly used to treat dairy animals and mastitis was treated with beta-lactam class of antibiotics. The prevalence of antibiotic residues in milk samples has been found to be higher in silo and tanker samples as compared to market and commercial pasteurized milk samples [19]. These findings prove that that antibiotic are used in dairy animals in these regions, though details of the frequency, duration, and reasons for use and overuse are not well recognized.

2.6 Use of antibiotics in poultry

The level of resistance in Indian poultry is reported to be high for many antibiotics. A recent study conducted by members of the Global Antibiotic Resistance Partnership [20] reported significant differences in the resistance pattern of broiler farms of Punjab with level of antibiotics used in normal poultry production. Results revealed that antibiotic use in broiler farms were likely to be more than 20 times to harbor-resistant *E. coli*, and prevalence of multi-drug resistance was much higher which was found 94% in broiler farms. In meat shops of Bikaner (Rajasthan), 96% of chicken samples contained *S. aureus* (n = 48), which were sensitive to ciprofloxacin, doxycycline, and gentamycin, and all were resistant to ampicillin, cloxacillin, and tetracycline [21].

2.7 Transfer of antimicrobial resistance from livestock to humans

Farm workers and slaughterers are at high risk of exposure to resistant antimicrobials due to direct contact with infected animals. Handling pigs and poultry while working in a farm environment puts farm workers at risk of picking up resistant bacteria from the animals' bodies or their feces. A study in the Netherlands in 2001–2002 revealed the same genetic patterns of resistance in *E. coli* samples from turkeys and broiler chickens, their farmers and slaughterers [22]. Consumption of food contaminated with resistant bacteria such as *Salmonella*, *Campylobacter*, and *E. coli* can increase the resistant bacteria in the human beings. Contamination of meat from fecal material getting onto the carcase during the slaughter and evisceration process, during the removal of animal gut, can contaminate other foods in domestic or restaurant/catering kitchens. The European Food Safety Authority (EFSA) revealed in 2010 that live chickens colonized with *Campylobacter* are 30 times more likely to contaminate meat as compared to uninfected birds [23]. Resistant bacteria can be transferred in water, soil, and air because animals excrete a significant amount of antibiotics they are administered, which make manure a potential source of both antibiotics and antibiotic-resistant bacteria that can enter soil and groundwater [15].

3. Rationale and approaches to limit the spread of antimicrobial resistance

Synchronization of international, national, and local approaches is advised for control and prevention of antimicrobial resistance. Promoting the rational use of antimicrobials, control on over-the-counter availability of antimicrobials, improvement of hygiene, prevention of infection, and control are the major recommended approaches. Thus, proper understanding of mechanism of resistance and accordingly innovation in development of new drugs is the need of the hour. A multidisciplinary, collaborative, regulatory approach is demanded for combating antimicrobial resistance [24].

3.1 Reasons to focus on antibiotics' use in livestock vis-à-vis antibiotic overuse and resistance

For decades, meat industry has fed antibiotics to chickens, pigs, and cattle for their weight gain and disease prevention in the stressful and unhygienic conditions that is prevalent in industrialized animal agriculture production facilities. A strong scientific consensus asserts that this practice fosters antibiotic resistance in bacteria, which is detrimental to human health (HSUS Report). Food animals are quite susceptible to benign or commensal opportunistic microbes, so they are often exposed to antimicrobials, such as the antibiotic, for disease treatment and prevention, sub-therapeutic purpose and prophylactic purpose to promote growth and improve feed efficiency. Many of these antimicrobials used to treat diseases common to both livestock species and humans closely resemble drugs used in this species [25]. On the one hand, these miraculous antimicrobial drugs are pillars of modern medicine to prevent and diagnose dangerous bacterial infections and save lives. On the other hand, the overuse, injudicious use, and misuse of these antimicrobial drugs have spawned the evolution of life-threatening bacteria that is making the current antibiotics reserve useless [26]. Thus, antimicrobial resistance can be defined as the ability of microbes, such as bacteria and fungi, to grow and continue to multiply even in the presence of administered antimicrobial with purpose to kill or limit their growth (NIAID).

3.2 Philosophy of judicious use of antimicrobials in line with animal welfare

Animal agriculture by human needs to be predicated on ethical judgments where sub- or non-therapeutic use of antibiotics on food producing animals on ethical judgement scale seems to be objectionable. The problem is that food-animal producers do not realize the ethics in their business because they claim that the conditions and processes in the factory farm are not a matter of ethics but of a societal necessity to fulfil the feed demand of the population. These producers seem to fail to realize the ethical dimensions of their practices, not only for food safety issues for consumers, but also welfare issues for their animals [27]. Any policy judgment including the danger of tolerable resistance or the level of animal abuse tolerable for the sake of the benefits from antibiotics overuse in animal feed is the subject of ethical judgment [28]. Lack of treatment protocols and solidarity of animal from herd and stopping the course of treatment after apparently realizing the disappearance also comes under the purview of animal welfare.

3.3 Current scenario in India and developing countries

Antibiotic use has been increasing steadily (e.g., between 2005 and 2009, 40% increase has been found in units of antibiotics sold). Cephalosporin sales increased by 60% over that 5-year period (in units sold) [20]. Antibiotics are used to treat human illness, livestock, and poultry diseases. In livestock sector, it accounts for more than 50% in order to control and treat diseases, and in low doses in animal feed, to promote growth and improve production of animal products [29]. There is no regulation in India to regulate the use of antibiotics in food animals, such as poultry and dairy animals raised for domestic consumption. As per, Prevention of Food Adulteration Rules (1995), Part XVIII: use of antibiotic and other pharmacologically active substances are applied only to certain types of seafood and poultry intended for export only [30]. Very few studies on antibiotic residues in animal products have been conducted in India, where one on honey was widely recognized [31]. Centre for Science and Environment, New Delhi, in a study revealed that 11 of 12 samples of honey taken from the domestic market were not in compliance with standards for its export. The level of antibiotic residues found was not high enough to cause an adverse effect in consumers, but it appealed for regulation and monitoring of antibiotic residues in honey because continuous long-term exposure to low levels of antibiotics could increase antibiotic resistance in pathogenic bacteria making their treatment difficult [32]. The National Policy for Containment of Antimicrobial Resistance—India was documented in years 2007, 2011, and 2017, which covers a range of topics, including reduction of antibiotic use in animals, particularly domesticated animals; provision of infection surveillance in hospitals; improving hospital surveillance for monitoring antibiotic resistance; promoting rational and judicious use of drug through education, monitoring, and supervision; researching new drugs; and developing and implementing a more restrictive and participatory antibiotic policy by including various stakeholders 65. Under the new Schedule H1, selling of antibiotics over-the-counter will be banned [20].

4. Delineating the use of antibiotics by farmers from farm-to-fork

Low income countries should follow the approaches of World Health Organization, World Organization for Animal Health, and the Food and Agriculture Organization of the United Nations, which recommends to implement national action plans encircling human, food animal, dairy animals, and environmental sectors to formulate appropriate policies, interventions, and activities that could address the prevention and containment of antimicrobial resistance from farm-to-fork. Suitable interventions should be designed, which include the following fields and coverage.

4.1 Use of antimicrobials by farmers

In strong sense, there is dependence among piggery farmers on antimicrobials to sustain production, improve farm performance, and maintain health status. Lack of concern about the harmful effects of antimicrobial use on their own and public health was identified among pig producers as a result of a reduction in the curative ability of antimicrobials and the selection of antimicrobial resistance bacteria [33]. A study conducted in Danish system revealed that 82% of antimicrobials sold by pharmacies were direct to individuals on prescription with specifications for use, 78% of antimicrobials sold by pharmacies used for pigs, and 20% for cattle [34]. The overuse of antibiotic has exploited this miracle drug to such an extent that a study in the Netherlands revealed that 79% of farmers used antibiotics routinely and 18% occasionally extended antibiotic treatment. The choice of progressive farmers for adopting prudent use of antibiotics by avoiding routine use of antibiotics was perceived as good practice by fellow farmers. This was followed by repeating the initial label treatments [35]. There are certain specific antibiotics which are used by farmers to treat animals without veterinary consultation (e.g., gentamicin in Ohio). Thus, improving information flow from Veterinarians to farmers may be the most effective means of promoting prudent use of antibiotics on dairy farms [36]. Subjective norms and moral obligations together, in which perceived moral obligations to peers, clients, and the regulatory norm setting sector associated with the feedlot industry increase social pressures to use antibiotics in acutely sick, chronically sick, and high-risk feedlot cattle [37].

4.2 Understanding the antimicrobials overuse in small dairy farms

The incidence of death of farm workers due to treatment failure attributed by antimicrobial resistance is likely much higher in developing countries where more people live in close contact with livestock, where food hygiene is not well practiced [38]. A report by WHO revealed that in developing countries throughout the world, even less than 50% of human are treated according to standard treatment protocol, and prescribing patterns were found substandard regardless of the type of prescriber [39]. Antibiotic-resistant food-borne infections, emergence of new multi-resistant strains of bacteria, and spread of resistant genes are some main areas of risk due to indiscriminate and overuse of antibiotics [15].

4.3 Antimicrobial resistance and intensive animal farming

The basic reason for increase in antimicrobial resistance in food animals is factory farming. In intensive pig and poultry production, animals are reared in confined and overcrowded conditions, usually with no outdoor access, and they are bred and managed for maximum production yield, i.e., to grow faster in size and number or to produce more meat, milk, and eggs. This forces them to compromise their health and their immune responses and encourage infectious disease to develop and spread easily among these livestock [40, 41]. Without the aid of drugs for disease prevention, it would not be possible to keep the animals productive in the intensive conditions, in which they are often kept and managed without proper care by the livestock keeper. Earlier, the policy-makers of 50 years ago permitted antibiotics to be used for non-therapeutic reasons in animal production, often in spite of scientific misgivings, which can be perceived as a serious mistake now. Fifty years later, while the evidence continues to be disputed by some sections of the industry, the actual and potential damage to public health is acknowledged by scientists and policy-makers due to the spread of antibiotic resistance among livestock species and human being (vertical and horizontal transfer of resistance) [15].

5. Synchronized efforts by stakeholders to reduce the pace of spread of antimicrobial resistance

All the stakeholders including veterinarians, paravets, farmers, and pharmaceutical companies should be made aware about their interacting roles from antimicrobial prescription to use, in which decision made by one stakeholders affects the worth and value of choices and decision for other stakeholder [42]. These aspects can be highlighted under following sub-heads.

5.1 One Health approach to combat antimicrobial resistance

One Health approach recognizes that human, animals, and ecosystem health are inextricably linked to each other. It came in to light because many factors have changed the interaction between humans, animals, and environment. Thus, for achieving the mutual optimal health outcomes, it needs the cooperation of human health, livestock, and environment health. Resistance to infectious diseases increase the cost of treatment as well as serious biosecurity concerns due to spread of antibiotic resistance. Thus, the animal production is hampered due to rise in incidence of infections.

5.2 Antimicrobial stewardship

The primary focus of an antimicrobial stewardship program is to optimize the use of antimicrobials to achieve the best treatment outcomes, reduce the risk of infections, reduce or stabilize levels of antibiotic resistance, and promote livestock safety. Creating an antimicrobial stewardship program needs baseline information, including institutional use of antimicrobial [43]. This would help to identify recurrent problems with antimicrobial use at the institution and frames the problems that need to be addressed [44]. The antimicrobial stewardship efforts should focus on improving adherence to documentation standards, optimizing the use of antimicrobials, appropriateness of drug dosing, halting treatment of asymptomatic bacteria and microbes, and minimizing the length of surgical prophylaxis [45].

5.3 Recommendations to control antibiotic overuse

Recognizing that antibiotic resistance is a reality crossing the geographical boundaries of the world, in developing countries, the prevalence of resistant microbes will rise over time, which demands urgent action. Vaccinations to prevent various disease falls into this category of recommendation, but their "antibiotic-sparing" effects are often overlooked because these are of secondary importance. Restricting the use of antimicrobials in livestock and poultry for non-therapeutic use, particularly growth promotion, could be beneficial. There is a need to eliminate irrational or inappropriate use, enforce prescription only laws, and eliminate over-the-counter antibiotic purchases, surveillance, distribution of Standard Treatment Guidelines (STGs), antibiotic sensitivity testing, checklists for surgical procedures, educating farmers and other stakeholders about appropriate use of antibiotics, and improving antibiotic supply chain and quality (Global Antibiotic Resistance Partnership (GARP)-India Working Group 2011). For gaining better understanding and subsequent action toward antimicrobial resistance, detailed social science research is needed to gather information on the processes of diagnosis, prescription, use of antimicrobials, the application of treatments besides antimicrobials, and the processes of data generation. Thus, sub-optimal use, potential users, and food chain pinch points could be identified. There is a general scantiness of data on on-farm application and use of antimicrobials. The tools for recording on-farm medicine use, such as paper spread sheets and computerized entries, may be of practical use to farmers in the health management of their animals/birds or to veterinarians in providing an accurate picture of how prescribed medicines are actually used [46].

5.4 Prescription of antimicrobials

Prescription of antibiotics are strongly influenced by the demand of farmers for antibiotics, fear of veterinarians blamed if antimicrobials later prove unnecessary, the expectation of farmers to be prescribed antimicrobials, confidence of veterinarians in diagnosis. Thus, prescription decisions are strongly influenced by multifactorial non-clinical influences such as farmer pressure and cost of drug, etc., to some extent [47]. Also, variations are present in beliefs of veterinarians regarding efficacy of systemic antibiotics for dry-cow therapy results in very different decisions being taken on farm and considerable discrepancies in treatment. Thus, it raises concern of the consistency and appropriateness of antibiotic prescription by them [48]. Antibiotic sensitivity testing should be preferred before prescribing the antibiotics [49].

6. Conclusion

The overuse of antimicrobials in livestock is leading to decline in antimicrobial effectiveness against infections in animals and eventually in humans. Use of antimicrobials purely as growth promoters and prophylactic purposes should be avoided and initiatives should be taken to phase out the sub-therapeutic use of antimicrobials. Injudicious use, overuse, and indiscriminate use of antimicrobials should be avoided. The obtaining of antibiotics from over—the counter sales should be checked and antimicrobial conservation practices should be encouraged to control the indiscriminate prescription and use of antimicrobials. Suitable strategies and policies should be formulated in line with the World Organisation for Animal Health and World Health Organization initiatives which call for harmonious efforts among stakeholders of different countries. Suitable extension outreach and continuing programmes should be devised to promote awareness among stakeholders about judicious use of antimicrobials and educate farmers, veterinarians, and consumers on the potential risk of antimicrobial resistance. There is need for surveillance and monitoring to track rates of antimicrobial use in veterinary sector, increase in resistance, and spread of antimicrobial residues in food chain.

Author details

Hans Ram Meena* and Vikash Kumar
Division of Dairy Extension, ICAR-National Dairy Research Institute, Karnal, India

*Address all correspondence to: drhrms@gmail.com

IntechOpen

© 2019 The Author(s). Licensee IntechOpen. This chapter is distributed under the terms of the Creative Commons Attribution License (http://creativecommons.org/licenses/by/3.0), which permits unrestricted use, distribution, and reproduction in any medium, provided the original work is properly cited.

References

[1] Hao H, Cheng G, Iqbal Z, Ai X, Hussain HI, Huang L, et al. Benefits and risks of antimicrobial use in food-producing animals. Frontiers in Microbiology. 2014;**5**:288

[2] Chuanchuen R, Pariyotorn N, Siriwattanachai K, Pagdepanichkit S, Srisanga S, Wannaprasat W, et al. Review of the Literature on Antimicrobial Resistance in Zoonotic Bacteria from Livestock in East, South and Southeast Asia. Bangkok, Thailand: Food and Agriculture Organization Regional Office for Asia and the Pacific; 2014

[3] World Bank Group. Drug Resistant Infections: A Threat to Our Economic Future. The World Bank 2016. 2016. Available from: http://pubdocs.worldbank.org/en/689381474641399486/1701381-AMR-Lab-Report-Web.pdf

[4] O'Neill J. Tackling a Global Health Crisis: Initial Steps. The Review on Antimicrobial Resistance. Chaired by Jim O'Neill. 2015. Available from: http://www.amr-review.org

[5] Anthony DS, Ramachandran R, David CL, Korinek A, Jillian PF, Heaney CD. A Framework for Costing the Lowering of Antimicrobial Use in Food Animal Production. Baltimore, USA: Johns Hopkins Center for a Livable Future; 2016

[6] Laxminarayan R, Van Boeckel T, Teillant A. The economic costs of withdrawing antimicrobial growth promoters from the livestock sector. OECD Food, Agriculture and Fisheries Papers, No. 78, OECD Publishing; 2015

[7] Bengtsson B, Greko C. Antibiotic resistance—Consequences for animal health, welfare, and food production. Upsala Journal of Medical Sciences. 2014;**119**(2):96-102

[8] Laxminarayan R, Van Boeckel T, Teillant A. Global antimicrobial use in the livestock sector. Organisation for Economic Co-operation and Development; TAD/CA/APM/WP(2014)34/FINAL; 2015

[9] World Health Organization. WHO Global Strategy for Containment of Antimicrobial Resistance (No. WHO/CDS/CSR/DRS/2001.2). 2001. Geneva: World Health Organization

[10] Laxminarayan R, Malani A, Howard D, Smith DL. Extending the Cure: Policy Responses to the Growing Threat of Antibiotic Resistance. Washington, DC: Resources for the Future. Earthscan; 2007

[11] Deshpande P, Rodrigues C, Shetty A, Kapadia F, Hedge A, Soman R. New Delhi Metallo-beta lactamase (NDM-1) in Enterobacteriaceae: Treatment options with carbapenems compromised. The Journal of the Association of Physicians of India. 2010;**58**:147-149

[12] Gelband H, Molly Miller P, Pant S, Gandra S, Levinson J, Barter D, et al. The state of the world's antibiotics 2015. Wound Healing Southern Africa. 2015;**8**(2):30-34

[13] CDDEP. Antibiotic Use and Resistance in Food Animals. Current Policy and Recommendations. Washington, DC, USA: Center for Disease Dynamics, Economics and Policy; 2016

[14] World Health Organization. Urgent Action Necessary to Safeguard Drug Treatments. 2011. World Health Day 2011, Press release 6 April 2011. Available form: http://www.who.int/

mediacentre/news/releases/2011/whd_20110406/en/index.html

[15] Turner J. Anonymous. Antibiotics in Farm Animal Production. Public Health and Animal Welfare. Compassions in World Farming. 2011. Available from: https://www.ciwf.org.uk/media/3758863/Antibiotics-in-Animal-Farming-Public-Health-and-Animal-Welfare.pdf

[16] Van Boeckel TP, Brower C, Gilbert M, Grenfell BT, Levin SA, Robinson TP, et al. Global trends in antimicrobial use in food animals. Proceedings of the National Academy of Sciences. 2015;**112**(18):5649-5654

[17] Ramakrishna Y, Singh RS. Residual streptomycin in milk. A survey. Indian Journal of Dairy Science. 1985;**38**:148-149

[18] Sudershan RV, Bhat RV. A survey on veterinary drug use and residues in milk in Hyderabad. Food Additives & Contaminants. 1995;**12**(5):645-650

[19] Unnikrishnan V, Bhavadassan MK, Nath BS, Ram C. Chemical residues and contaminants in milk: A review. The Indian Journal of Animal Sciences. 2005;**75**(5):592-598

[20] Global Antibiotic Resistance Partnership. Rationalizing antibiotic use to limit antibiotic resistance in India. The Indian Journal of Medical Research. 2011;**134**(3):281-294

[21] Hemlata. Enumeration and antibiotic resistance pattern of *Staphylococcus aureus* from raw chicken meat sold in Bikaner city. Journal of Pure and Applied Microbiology. 2015;**9**(2):1725-1730

[22] Van den Bogaard AE, London N, Driessen CAGG, Stobberingh EE. Antibiotic resistance of faecal Escherichia coli in poultry, poultry farmers and poultry slaughterers. Journal of Antimicrobial Chemotherapy. 2001;**47**(6):763-771

[23] European Food Safety Authority. (EFSA). Analysis of the baseline survey on the prevalence of *Campylobacter* in broiler batches and of *Campylobacter* and *Salmonella* on broiler carcasses, in the EU, 2008-part B: Analysis of factors associated with *Campylobacter* colonisation of broiler batches and with *Campylobacter* contamination of broiler carcasses; and investigation of the culture method diagnostic characteristics used to analyse broiler carcass samples. EFSA Journal. 2010;**8**(8):1522

[24] Uchil RR, Kohli GS, KateKhaye VM, Swami OC. Strategies to combat antimicrobial resistance. Journal of Clinical and Diagnostic Research. 2014;**8**(7):ME01

[25] McEwen SA, Fedorka-Cray PJ. Antimicrobial use and resistance in animals. Clinical Infectious Diseases. 2002;**34**(Supplement_3):S93-S106

[26] Levy SB. Before the Subcommitte on Health of the U.S. House Committee on Energy and Commerce (Tufts University School of Medicine, 14 July 2010). 2010. Available from: http://emerald.tufts.edu/med/apua/policy/7.14.10.pdf

[27] Levi A. Why we should talk about animals when we talk about antibiotics [Doctoral dissertation]. Colorado, United States: Colorado State University, Libraries; 2016

[28] Rollin B. Ethics, science, and antimicrobial resistance. Journal of Agricultural and Environmental Ethics. 2001;**14**(1):29-37

[29] Manna SK, Brahmane MP, Manna C, Batabyal K, Das R. Occurrence, virulence characteristics and antimicrobial

resistance of *Escherichia coli* O157 in slaughtered cattle and diarrhoeic calves in West Bengal, India. Letters in Applied Microbiology. 2006;**43**(4):405-409

[30] Export Inspection Council of India. Approval and Monitoring of Processing/Storing Establishments for Export: Egg Products. New Delhi; 2007. Available from: http://www.eicindia.org/eic/inspection/egg.pdf

[31] Anonymous. Antibiotics in most honey brands: Study. 2007. The Times of India. Times News Network. September 16, 2010. Available from: http://articles.timesofndia.indiatimes.com/2010-09-16/india/28225876_1_honey-brands-antibiotics-honey-samples

[32] Johnson S, Jadon N, Mathur HB, Agarwal HC. Antibiotic residues in honey. Report September (2010) Centre for Science and Environment, New Delhi, India; 2010

[33] Moreno MA. Opinions of Spanish pig producers on the role, the level and the risk to public health of antimicrobial use in pigs. Research in Veterinary Science. 2014;**97**(1):26-31

[34] Stege H, Bager F, Jacobsen E, Thougaard A. VETSTAT—The Danish system for surveillance of the veterinary use of drugs for production animals. Preventive Veterinary Medicine. 2003;**57**(3):105-115

[35] Swinkels JM, Hilkens A, Zoche-Golob V, Krömker V, Buddiger M, Jansen J, et al. Social influences on the duration of antibiotic treatment of clinical mastitis in dairy cows. Journal of Dairy Science. 2015;**98**(4):2369-2380

[36] Cattaneo AA, Wilson R, Doohan D, LeJeune JT. Bovine veterinarians' knowledge, beliefs, and practices regarding antibiotic resistance on Ohio dairy farms. Journal of Dairy Science. 2009;**92**(7):3494-3502

[37] Jan JS, McIntosh W, Scott HM, Dean W. The effects of moral obligations to others and others' influence on veterinarians' attitudes toward and recommendations to utilize antibiotics in feedlot cattle. Journal of Rural Social Sciences. 2010;**25**(2):24-33

[38] Lanata CF. Studies of food hygiene and diarrhoeal disease. International Journal of Environmental Health Research. 2003;**13**(sup1):S175-S183

[39] Medicines WE. Pharmaceutical Policies: Medicines Use in Primary Care in Developing and Transitional Countries: Fact Book Summarizing Results from Studies Reported between 1990 and 2006. In: Holloway K, Ivanovska V, Vialle-Valentin C, Johnson A, Lewis S, Wagner A, Ross-Degnan D, editors. Geneva: World Health Organization; 2009

[40] Greger M. The human/animal interface: Emergence and resurgence of zoonotic infectious diseases. Critical Reviews in Microbiology. 2007;**33**(4):243-299

[41] Humphrey T. Are happy chickens safer chickens? Poultry welfare and disease susceptibility. British Poultry Science. 2006;**47**(4):379-391

[42] Kumar V, Gupta J. Synthesis of policies among stakeholders to combat antimicrobial resistance in livestock animals: Indian perspective. Journal of Pharmacognosy and Phytochemistry. 2018;**7**(2):3649-3653

[43] Drew RH. Antimicrobial stewardship programs: How to start and steer a successful program. Journal of Managed Care Pharmacy. 2009;**15** (2 Supp A):18-23

[44] Tamma PD, Holmes A, Ashley ED. Antimicrobial stewardship: Another focus for patient safety? Current Opinion in Infectious Diseases. 2014;**27**(4):348-355

[45] Leung V, Li M, Wu JHC, Langford B, Zvonar R, Powis J, et al. Evaluating antimicrobial use and spectrum of activity in Ontario hospitals: Feasibility of a multicentered point prevalence study. Open Forum Infectious Diseases. 2018;**5**(6):ofy110

[46] Buller H, Hinchliffe S, Hockenhull J, Barrett D, Reyher K, Butterworth A, et al. Systematic review and social research to further understanding of current practice in the context of using antimicrobials in livestock farming and to inform appropriate interventions to reduce antimicrobial resistance within the livestock sector. DEFRA Report; 2015

[47] Gibbons JF, Boland F, Buckley JF, Butler F, Egan J, Fanning S, et al. Influences on antimicrobial prescribing behaviour of veterinary practitioners in cattle practice in Ireland. Veterinary Record. 2013;**172**(1):14

[48] Higgins HM, Dryden IL, Green MJ. A Bayesian elicitation of veterinary beliefs regarding systemic dry cow therapy: Variation and importance for clinical trial design. Preventive Veterinary Medicine. 2012;**106**(2):87-96

[49] Coyne LA, Pinchbeck GL, Williams NJ, Smith RF, Dawson S, Pearson RB, et al. Understanding antimicrobial use and prescribing behaviours by pig veterinary surgeons and farmers: A qualitative study. Veterinary Record. 2014;**175**(23):593-593

Chapter 3

Farm Animals' Health Behaviours: An Essential Communicative Signal for Farmers' Veterinary Care and Sustainable Production

Okanlade Adesokan Lawal-Adebowale

Abstract

Farm animals constitute valuable source of quality nutrition and economic development across the world, and sustainable farm animal production is greatly being challenged by pests and disease infestation with the resultant poor productivity, death of animals and economic losses to the farmers and nations at large. But before infections reached the threshold of debilitative effects, an infected animal communicates a physiological disturbance by vocalisation and/or visual cues. While a healthy animal communicates its good health status by active display and movement of the body parts in response to its environment, a sick animal manifests its health situation by looking dull, by being self-isolated from the stock, by being sluggish or by refusing to move on when approached or to be fed. Although the communicated cues by a farm animal are determined by the kind of physiological impairment experienced by the animal, farmers' understanding of the specific communication cues by the farm animals would make quick detection of any laden disease in the animals possible and stimulate prompt health care service provision. Consequently, several ways by which farm animals communicate their health situation and the veterinarian actions to be taken in the light of a disease outbreak are highlighted in this article.

Keywords: farm animals, pests and diseases, animal behaviour, animal health communication, farmers' cognition of animal communication, veterinary care

1. Introduction

Farm animals constitute valuable bio-resources that support man's social and nutritional security through provision of quality food resources, specifically in the form of meat, milk, egg, and other food-based by-products, and as well economic development of human society through management and processing of the animals into products and by-products for income generation by individuals and organisations. In this wise, the livestock sector makes valuable contributions to national gross domestic products (GDP) of countries around the world. It contributes about 40% of the global value of agricultural outputs, supports the livelihood and food security of almost 1.3 billion people, and offer opportunities for agricultural development, poverty reduction and food security globally [1]. Animal-based

IntechOpen

foods provide a variety of micronutrients that are essentially deficient in plant-based foods [2]. While crop-based foods are deficient in vitamin A, vitamin B-12, riboflavin, calcium, iron and zinc, animal-based foods are particularly rich in these nutrients thereby serving as food resource with which multiple nutrient deficiencies and nutrition challenges that arise from heavy consumption of crop-based foods can be readily addressed [3]. The effects of such animal-based food include good growth, better weight gain and healthy condition, particularly in children [4–6].

In addition to the nutritional values of farm animals is the stock's contribution to economic development around the world. The animals not only serve as means of employment to the farmers but also serve as means of income generation through direct sales of live animals and products such as meat, milk, eggs, wool and hides. Among all other animal products, sales of milk and eggs are essentially a means of continuous cash flow by which farm families move from subsistence to cash-based economies [7] and as well meet other essentialities of life. Similarly, animal husbandry stimulates the development of animal-based food industries through provision of animals and animal products as production resources for processing and conversion into other by-products or animal-source foods. The animal-based food industries not only become a source of income generation owner of the industry, through the distribution and marketing of the produced animal-source foods, but also become a job-creating avenue in the context of production, administration, financial management, marketing and distribution along the production chain to guarantee employment opportunity for individuals seeking to work. The economic values of the livestock sector are however not only limited to the immediate environment (local areas) of production but also have much impact on regional, national and international economies. For instance, the livestock industries contribute a lot to the United States of America's (USA) national economy by providing about 1,851,000 jobs, contributing $346 billion as total economic outputs, $60 billion as household income, $15 billion as income from paid taxes, and $6 billion from property taxes [8]. In the region of South Dakota, where livestock is dominantly managed in the Unites States, the livestock sector provides as much as 29,020 jobs, contributing $7.3 billion in total economic outputs, $1.1 billion in household income, $235 million in income taxes paid, and $149 million in property taxes paid [8].

In Europe. The livestock sector contributes almost 125 billion Euros per year and accounts for 40% of total agricultural production in the area [9]. Also, the contribution of animal production to the gross indigenous production in the region of Europe is about twice as high as the whole agribusiness sector in the area. In the developing countries, livestock contributes about one-third of the total agricultural outputs [10] and 10–45% of GDP in the region. In Ethiopia, the sector contributes an estimate of 19% of the country's total GDP, 45% of the agricultural GDP, and about 20% of the country's export earnings [11]. Based on the social and economic values of livestock a great deal of efforts in terms of research and management has been jeered toward improved and sustainable livestock production around the world.

2. Livestock development to improve productivity

In an attempt to sustain the social and economic values of the farm animals for nutrition and economic benefits of man, research has brought about a great deal of improvement in the livestock sector both in terms of production and management on a global scale. With the USA and Japan's support for livestock development, livestock production technology has developed rapidly worldwide [9]. The development strategies are generally in the form of breeding/cross-breeding, quality feed development, improved healthcare services, animal husbandry, and marketing system. Also is the development of new or automated technologies to aid the management and well-being of the animals [9]. Impacts of research in breeding include the development of quality breeds of animals with the potential for improved productivity, quick maturing, prolific litter production and resistance or adaptability to changing the environment. In this regard is the emergence of breeds of farm animals with specific production potentials in terms of meat, milk, and eggs. For instance, poultry birds such as broiler—a fast-growing chicken, is specifically developed for meat production while layers are developed for egg production. Pigs are largely developed for pork or beacon (meat) production and given the prolific litters produced at a birth, the animal has been a good source of meat for nutritional security of human society. Alongside the monogastric farm animals is the ruminant comprising sheep, goats, and cattle.

Cattle breeds such as Ayrshire, Brown Swiss, Friesian, etc. are prolific in milk production while the likes of Simmental, Angus, Charolais, Hereford, are good meat (beef) producing breeds. In the same vein are breeds of sheep, such as Dorper, Hampshire, Dorset, Suffolk; and goats—Boer, Spanish, Range Land, Kalhari, etc. with good meat (mutton and chevon, respectively) production. Although milk production is largely from dairy cattle, breeds of sheep—East Friesian, Lacaune, Finnish Landrace Polypay; and goats—Alpines, LaMancha, Saanens, Nubian, and Nigerian Dwarf, etc. are good milk-producing small ruminants [12, 13]. In the same vein are breeds of sheep such as Merino, Leicester Long-wool, Lincoln; developed for wool production alongside the likes of Barbados, Jacob; developed to produce hair for industrial fabric production. The developed prolific breeds of sheep for large litters of birth include Finnsheep, Romanov, and Booroola Merino, etc. [12, 13] thereby providing the sheep keepers the opportunity for quick increase of their flock population within a short time. In the same vein, pig prolificacy brings about 12–14 piglets per litter thereby making the animal the most prolific stock for production by intending livestock farmers.

To support the productivity of the developed farm animals is the development of quality feeds that could adequately supply the animals' energy requirement for production and maintenance. Although feed types of the various farm animals differ, the common denominator of them all is the nutritional contents of the feed, characterised by the presence of adequate crude protein, carbohydrates, fatty acids, minerals, and vitamins; that may be available to the animals from the given feeds. In light of this are the development of different feed formulae and automated feed processing technologies for the production of feeds that are adequately rich in nutritional contents. In addition is production of concentrates and other feed additives to enhance feed intake and nutrition of the animals for improve productivity [10] An optimal nutritional program ensures adequate intakes of amino acids (both essential and nonessential), carbohydrates, fatty acids, minerals, and vitamins by animals such that the nutrition contents readily supplement deficiencies in basal diets of the animals for enhanced consumption and conversion [14]. In addition to supporting productivity of farm animals quality feed is of significant value to healthy leaving of the animals as this readily enhance key metabolic functions of the animals to improve fertility and reproductive efficiency, immune function and animals' response to vaccinations, neonatal survival, and growth, feed utilisation efficiency, and meat quality [15]. In line with this is the development of health technologies and human capacities for the detection and effective management of farm animals. Deployment of automated technologies in farm animal management has widely been used, not only to monitor feeding behaviour and intake of animals but for detection ill health in the stock [9]. Consequently, well-fed and healthy animals result in good economic returns for the farmers in terms of profitable production arising from fewer expenses on animal medication, improved productivity and good marketing price.

3. Sustainable livestock production and the challenge of animal pests and diseases

With development efforts in the livestock sector, world animal production has continued growing with increased productivity. Between 1995 and 2002, the world total meat production increased by 19%, the total milk production by 11%, and the egg production by 23% [9]. Notwithstanding the research support for the development and improved productivity of the livestock sector, the sustainability of livestock production is still being challenged by several production and environmental factors; with the degree of impact being influenced by the management system put in place by farmers. While farmers may try as much as possible to give their animals all the necessary care in terms of good housing units—spacious, dry and well ventilated, quality feeds and water, and necessary healthcare, it remains inevitable to completely prevent the incidence of pests and diseases in livestock management basically because of the impossibility of eradicating disease-causing pathogens which could survive in almost all avenue. Reduction of the incidence of pests and diseases in farm animal management has though been achieved by a combination of good hygiene, appropriate use of vaccines and medical therapy, and selection of disease-free breeds of livestock, the result is by no means an eradication of infectious and contagious diseases owing to the fact that a multiplicity of pathogenic agents are inducing emergence of complex diseases that may even be difficult to diagnose [16].

Pathogens, which are groups of organisms that cause diseases in farm animals, exist in form of viruses, bacteria, fungi, protozoa, and parasites, and are so small that they cannot be ordinarily seen by sight except with the aid of electronic magnifiers. These pathogenic micro-organisms or microbes exist in different sizes and shapes and are readily transmitted to animals through various substrates or vectors such as water, soil, waste or faecal matter, humans and animals [17–19]. Viruses are very small micro-organisms and much smaller than all other pathogenic organisms such that it can only be viewed with a strong electronic microscope [16]. However, the micro-organisms can only survive and/or multiply within living cells basically because it lacks cell membranes, cytoplasm, ribosomes, and other cell organelles [20, 21]. This mode of survival and multiplication by viruses make it possible to destroy the cells of the infected host thereby result in certain diseases. About 60% of animal diseases are caused by a virus and unfortunately, such viral diseases have no cure thereby resulting in the death of infected animals [17]. On this note, preventive actions, usually by vaccination, are required to ensure the healthiness of farm animals. Viral diseases of farm animals include foot and mouth disease, rinderpest, bluetongue, vesicular stomatitis, swine fever, fowl pox, avian influencer, etc.

Bacteria, on the other hand, are relatively larger than viruses and also take different shapes and sizes but most are visible under an ordinary microscope. The micro-organisms, unlike viruses, could survive anywhere—within and outside living cells but could sporulate to form a protective coat that makes them survive in any environment for a long period, even years, and later cause infectious disease(s) in farm animals [16]. Bacteria that survive outside living cells could remain inactive until the emergence of favourable conditions to gain entry into the body of the animals, either through the skin or the eyes, breath into the lungs, consumed through food and water, to cause infections [22]. Similarly, some bacteria within the living cells, such as *Bacillus* and *Clostridium* species, envelope themselves with protective spores or endospores, which become dehydrated and highly resistant to an environmental condition such as heath, cold, or chemical compounds. The endospores, within which the bacteria may remain inactive for many years, produce endotoxin—a deadly substance that causes disease in animals [23, 23]. Bacterial diseases of farm animals include botulism, paratyphoid, anthrax, brucellosis, foot rot, tetanus, etc. Bacterial diseases are however not as deadly as viruses as these could be readily treated with antibiotics, particularly when detected on time. Besides, not all bacteria cause disease as some are of great value in ruminant digestion where they aid fermentation of consumed herbage in the rumen [24].

In similarity to bacteria, fungi are widespread, exist in various shapes and sizes, and could survive in nearly all forms of environment, be it water, soil, air or in the mould on stale food and mushroom for a long period that runs into years. The micro-organisms are though ordinarily harmless, they cause disease in some situations which can be damaging or even devastating in some cases [17, 25, 26]. Some fungal diseases affect the mucous membrane (mucosal) though, most of them affecting the skin (cutaneous) by colonising and destroying their tissues [27]. Skin diseases are though rarely critical or deadly, their resultant irritation discomfort the animals leading to inhibition or disruption of their feed intake and consequently, a drastic drop in productivity [28]. Diseases of fungi could be highly contagious and primarily spread by direct contact between animals; with the clinically infected animals as the greatest source of infections [26]. Examples of fungal diseases include ringworm, aspergillosis, candidosis, mycotic, protothecosis, dermatophytes, etc.

Protozoa are single-cell microbes that survive both inside and outside living cells and could be found in most habitats. Most of the protozoa are though harmless and even play a vital role in controlling bacteria population and biomass; some are however parasitic pathogens of humans and animals [29] and as such significantly cause diseases in farm animals and even as potential drivers of zoonotic transmission [30, 31]. The parasites are a significant cause of abortion and infertility in domestic ruminants [32] and a relatively uncommon group of respiratory ailment [33]. Epidemiological situations of protozoan infection may occur as a single infection or zoonoses and may be sporadic in otherwise healthy hosts [34]. Also, infection of the nervous system is mostly fatal. An epidemiological situation that corresponds to pseudoepidemics occurs in a large host of the animal population due to a common source or poor housing conditions of the animals characterised by poor-quality or badly-stored bedding. Protozoan parasites in poultry are coccidia (species of the *Eimeria* genus), cryptosporidia (*Cryptosporidium baileyi*), and histomonads (*H. melegridis*).

Alongside the pathogenic micro-organisms causing diseases in farm animals are parasites that depend on the animals for growth and/or survival [34, 35]. The parasites may live and survive inside or outside the body of farm animals as internal and external parasites. Common parasites of farm animals exist as worms, flukes, protozoa, and insects such as lice, mites, ticks, flies. Unlike the parasitic protozoa, most parasites are visible to sight but some mites and worms can only be seen under a microscope at their early stages [16, 34]. With the dependence of the parasites on host animal(s) for survival, they may either inhibit the normal physiological functioning of the animals or act as a vector of other diseases thereby resulting in negative impacts on the health and welfare of animals. Such negative impacts, particularly by the internal parasites, may be manifested as anaemia as a result of substantial blood loss, reduction of the animals' appetite resulting in debilitating health and susceptibility of the animals to other diseases, diarrhoea and death of the animals arising from severity of the parasitic impacts [34, 36, 37]. The external parasites, on the other hand, cause open sores on the skin of livestock which becomes irritating and annoying to the animals thereby causing them to reduce grazing and feed consumption.

4. Animal health behaviours: the communication signals for farmers' attention

One of the managerial goals of livestock keepers is the maintenance of good health of their animals as this is crucial to achieving profitable and sustainable animal production. On this note, livestock farmers try as much as possible to keep the animals free of infestation and infections, through hygienic practice and possibly vaccination of the animals against certain disruptive or deadly diseases. However, the ubiquity of pathogenic microbes (protozoa, bacteria, viruses, fungi, parasites) and other external parasites in a production environment of farm animals ultimately prone the animals to infections or infestation of pests and diseases which on gestation may turn out to be chronic or deadly [16]. Consequences of this are poor productivity by animals, increased cost of animal production arising from treatment or veterinary services, economic loss of animals, hindrance of production and/or productivity of animal food source industries, and possibly impairment of human health by infections from the animals [15]. In light of the economic implication of animal diseases, conscious actions need to be taken by farmers to prevent or promptly control any emerging disease of animals in their stock. However, before an infectious disease in farm animals begin to manifest its symptoms or get to the threshold of economic losses, the infected or physiologically disturbed animals ordinarily communicate their health status for the attention of their keepers.

By nature, animals ordinarily communicate with conspecifics or fellow animals in four basic ways, namely pheromones, auditory, visual and tactile cues [38–40]; they however indirectly communicate their social and health statuses to their keepers particularly using the auditory and visual cues. This is based on the fact that both animals and man could make and receive sounds, which are an essential stimulus to effecting responses between the two Animalia. In this wise, farm animals use their vocal sounds to express their health or social conditions to the farmers. For instance, a distressed animal makes distressing vocalisation as a way to call its owner for needed attention. Empirical studies on emotional vocalisations of farm animals [41–45] revealed that environmental stimulus and/or hormone concentrations affecting the mood, thirst, and hunger, and appetitive behaviour of an animal stimulate specific behaviours that may be accompanied vocalisation in the animals [41]. Thus, a dam in parturition distress might make a very high pitch sound to attract an attendant to give needed help for safe delivery; or where its kid is hooked and needed help to have it rescued. In the same vein, ewes or nannies on heat give constant high pitch sound as a way to indicate readiness for a reception which a breeder needs to take advantage of either by the introduction of ram/billy or artificial insemination. Pig is known for screaming when put under stress, particularly when being forcefully pulled in an attempt to move it from one place to the other. This cry may constitute a security call to the owner when the animal is to be forcefully taken out of the herds by an intruder.

The visual cues, on the other hand, are displayed actions that could be visually perceived by the farmers in their animals. Farm animals thus express their social and health situations by visual communication signals which a farm attendant needs to understand for an appropriate response. For instance, healthy farm animals are ordinarily active and ever ready to feed, and will be on the rise when they are approached. Where a farm animal sluggishly or refuses to move, or is reluctant to get up when approached (**Figure 1**) suggests that something is wrong with the animal and as such will need to be attended to for detection a laden health issue. Although, varying diseases have different physiological effects on specific farm animals, infected animals react on whatever kind of disease that might impair their physiological status

Figure 1.
Healthy lambs on their feet with the physiological lamb siting isolated and unable to move. Source: Photo by author.

by the display of signals that serve as health communication[1]. Common communicative signals by a disease-laden animal could thus take the following forms:

a. **Poor appearance of animals:** ordinarily, a healthy animal will generally have a good look with a characteristic good body structure and posture. Related body indicators of animals with good health include robust outlook and/or roundness of the stomach, smooth and well-laid hair or wool with uniform coverage over the body, bright eyes, and dry nose and mouth. On the other hand, a sick animal will put up a poor appearance that may be characterised by leanness, caved-in stomach, falling air or wool, nasal discharge, dripping salivation and/or continuous coughing. The caved-in stomach may have to do with poor feeding or poor feed conversion efficiency by the animal, while the falling hair or wool may be due to ectoparasite infestation of the animal. For instance, piglets severely affected by *coccidia* produce a rough-haired coat, become dehydrated and remain continually dirtied with faeces [46]. In the case of poultry, appearance of a bird with ruffled feathers instead of smooth and glossy ones or the wattle looks dull with lesions instead of being bright suggests impairment of the bird's physiological functions and as such will need veterinary care. A common disease of poultry with these characteristic features is the blackhead disease.

4.1 Inactive socialisation

A healthy animal is socially active and ever alert to its environment by having its head raised in an attempt to keep watch of its surroundings, and usually in close groups with one another. In ruminants, the animals are constantly chewing due to regurgitation of consumed pasture. *Babesiosis* disease, which is common to cattle, is known to cause cessation of rumination or constipation in the animal [47, 48]. Poultry birds in cages, deep litters, and free-range will equally have their heads up clucking or gobbling (respectively to chicken and turkeys) which increases on sighting someone or something strange in their environment. Where an animal is in isolation of the other animals or has its head lowered or drooped down and becomes dulled, unable to stand up or move sluggishly when being approached imply that the animal is physiologically disturbed and as such will need an examination to detect what is wrong with it.

[1] There abound diseases of farm animals with specific causative agents, symptoms, treatment, methods of control and prevention which is beyond the scope of this chapter. This chapter has only highlighted possible signs that an animal might display as an indicator of ill health which could be readily perceived by livestock farmers for prompt veterinarian actions.

4.2 Wobbled movement or gait

Whenever there is a need for farm animals to move, they will move steadily and easily; and where there is the need to flee from threats in their environment, they move very fast and run. Where an animal moves sluggishly or could not move would imply ill health. Abnormal gait can include unusual walking patterns or uneven weight-bearing, as seen when a cow is suffering from lameness [49]. Lameness may be an indication of rot in the foot characterised by swelling and moistened skin between the claws and foul-smelling discharge. Consequently, the animal remains lying down for long periods and may not bear weight on the affected leg, and where both front legs are affected, sheep, for example, walk on their knees and severe cases and chronic infection leads to grossly misshapen and overgrown hooves [50]. In the same vein is an abnormal stance indicating pain and this may be reflected as tucked abdomen and tail, hunched back or standing still for extended periods [49].

4.3 Drop or refusal of food consumption

Animals are ever ready to feed when nothing is wrong with them and as such a drop in the rate of feed consumption or outright refusal of feed by an animal or non-excitation at being fed would imply that something is wrong with the animal. Most diseases though cause an animal to refuse feed, nutritional factors equally accounted for feed refusal. For instance, diseases such as bloating, grass tetany, ketosis, hypocalcaemia, and mineral deficiencies are caused by nutrient deficiencies, excesses or imbalances, or by metabolic disturbances [51]. As a result of poor feeding or feed conversion efficiency, the animals lose weight with characteristic caved-in stomach and general weakness.

4.4 Abnormal droppings/dungs

Excreta of farm animals should be firm or looks 'bolus' and black or darkish green, particularly in cattle and pigs, and like small balls in sheep and goats. Droppings in poultry are usually greyish with urinary liquid. Where the dung of an animal looks watery, and sometimes have the faeces stained with blood would imply that something is wrong with the animal. Such signs usually have to with diarrhoea or other gastrointestinal diseases. Coccidia in piglets, for instance, is characterised by diarrhoea and scour in early stage, and late-stage, faeces become yellow or creamy-grey diarrhoea causing loss of condition and reduced growth rates at age 7–10 days [45]. Mortality rates may reach 20% [52]. Concurrent infections with other bacteria, viruses or parasites can increase mortality further.

4.5 Abortion in animals

One of the production goals in livestock management is a production of litters either for multiplication of animal population or marketing for income generation. A farmer could readily achieve this with healthy animals but where a pregnant animal is infected with the abortion-related disease the resultant effect is loss of foetus. Diseases such as leptospirosis, vibriosis, pestivirus, and trichomoniasis cause abortion which may be early-term abortion or embryonic loss in ruminant farm animals [51]. Also, there may be stillbirth, weak, stunted or deformed calves, and low calving and lambing rate. Observation of these traits in the animals calls for veterinary care of the animal(s).

4.6 Drop-in productivity

A careful look at production records of the animals could help determine the health status of the animals. An animal that has been prolific in production certainly loses production potential when its normal healthy condition is hindered by a disease or pest infestation. In a lactating dam with impaired physiological functions, there may be a drop in the quantity of produced per day or drop in egg production by sick birds [53]. Dressing of diseased animals also produces poor quality meat that is unfit for consumption thereby leading to loss of revenue or profitable income to the farmer.

4.7 Sudden death of animal(s)

The end of physiologically disturbed animal(s) is death, especially where the signs of ill health are not quickly detected for prompt veterinary action. In some cases, however, death may be sudden without a physical sign of ill health. Several diseases are so virulent that it leads to the death of farm animals within a short time. For instance, anthrax could cause sudden death within 2–3 days in ruminant and pigs. In the vein, PPR (Peste des petits ruminants) in small ruminants is associated with high morbidity and mortality [54, 55].

5. Economic implications of animal diseases

Diseases of farm animals generally cause a lot of losses either directly or indirectly in livestock production. Direct economic implications include decreased productivity of the animals, cost of disease control of animals, death of farm animals and loss of means of livelihood or economies of farmers. A diseased animal generally has its physiological functions impaired and as such will not be able to perform maximally or optimally. In the light, production outputs of farm animals in terms of egg, milk, meat, and wool production gradually decline [56] thereby resulting in inefficient production of the farmers. In other words, what farmers get as returns from production is far less than what is invested in the production of the animals. This is usually due to decreased food intake and/or inability of the animals to efficiently convert consumed feed into metabolic energy required for productivity. Alongside this is loss of quality products—milk and meat that might come from infected animals thereby losing market acceptability or value [53]. In severe cases and depending on the virulence of the disease, animals might lose weight appreciatively and die. This incur great loss to farmers as opportunity for production of kids is lost due to disease-related abortions by dams, revenue that could have accrued from sales of animals and animal products, and ultimate loss of means of livelihood in case of high mortality or death of herds and flocks which may be difficult to replace in most cases. An attempt to save the stock and sustain production result in increased cost of production arising from the cost of drug acquisition for treatment of the animals and of veterinary services. Economic losses to the world poultry industry are believed to be more than US$3 billion annually [57]. Back in the year 2000 in rural communities in the northern part of Cameroon, an estimated value of sheep and goats losses to PPR was put at US$53, 902 over five years, while in Bangladesh, the estimated value of goats mortality from PPR was put at US$34.8million in the year 2001 [56]. Also is a devastating effect on cash flow and equity at the farm or industry level arising from sharp fall in consumers' demand for primary products of livestock out of concern of zoonosis, and severe limitation or elimination of animal marketing options [18, 55]. In the same vein, access to the premium export market is affected as animal-source food from a region with disease outbreak will not be accepted in the international markets thereby distorting the development of the livestock sector both within the country(ies) in question and globally [56].

Beyond the farm, level is indirect consequences of disease in animals which include zoonotic and cost of human treatment, loss of employment to farm labours and employment in other livestock values chains or industries, alteration of nutritional balance of the populace, change in consumer behaviours and marketing shocks. Humans become infected with animal diseases either from consumed products of infected animals or the circulation of zoonotic agents between animals, humans, and the environment with hindrances of their wellbeing and economic activities. The direct cost of zoonotic diseases over the last decade is estimated to be more than $20 billion with over $200 billion indirect losses to affected economies as a whole [58]. With marketing shocks, characterised by either short supplies of animals and animal products and/or an increase in prices of the available ones, consumers to make rational decisions to opt for alternative safe and cheap food sources. For instance, short supply of livestock products such milk, eggs and poultry meat, which tend to be cheapest, may lead to increase in prices thereby forcing consumers, particularly the poor people, to substitute with vegetable proteins or consume more of carbohydrate food sources with consequential less balanced diet [56]. In addition to related diseases of farm animals' disruption of nutrition security of the human populace is the transmission of such disease(s) to human populace either directly from the animal or on consummation of food products from the diseased animals. Farmworkers are most affected directly due to contacts with animals they care for and indirectly with the general populace through the consumption of disease-laden animal products. As expressed by the world society for the protection of animals, as much or more than 500 different pathogens, be it viruses, bacteria or parasites, are be transmitted from animals to humans through contact with live animals and dust inhalation, and consumption of disease-laden meat and animal by-products [59]. With numerous cases of animal diseases, particularly Avian Flu, Swine Flu, infecting people as well as the workers and veterinarians, the spread of animal viruses to humans is thus a serious public health concern and as such, they need for control and prevention of diseases in farm animals.

6. Farmers' cognition of farm animal-health behavioural communication and the veterinary responses

Given the social and economic implications of diseases of farm animals, it is essential that veterinary actions are taken to prevent, control and treat animals of any emerging disease. It is however of great value to have earlier detection of emerging or laden diseases where it is invariably impossible to prevent disease outbreak as this will save a great deal of the cost of treatment and prevent possible loss of animal to death. To achieve this, it becomes essential that livestock farmers have good cognition of their animal communication. The concept of animal communicate is generally grounded in animal behaviours whereby animals' social actions are interpreted with meaning either by fellow animals or humans. Consequently, by behavioural actions, animal communication is described as the process by which an animal transmits information to other animals (**Figure 2**) for incorporate into their decision making [60] or cause some kind of change in the animals that gets the information [61]. The transmitted information by animals in the communication process is however in form signals often reflected in sounds, colour patterns, postures, movements, electrical discharges, touches, the release of odorants, or

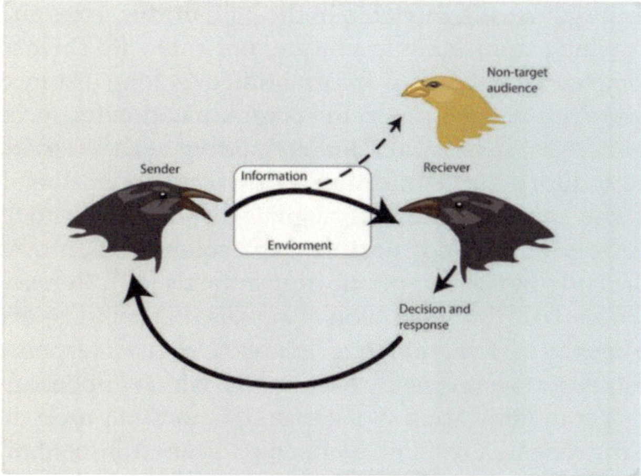

Figure 2.
A model of communication flow between animals. Source: Gillam [62].

Figure 3.
Vocalisation by a bird to its environment for a specific purpose. Source: Khan Academy [63].

some combination of these mediums [60]. These series of signals of animal communication are thus classified into four basic categories, namely visual, auditory, tactile and pheromone cues [60].

The auditory communication cue entails the use of the vocal cord (**Figure 3**) for sending sounds or cries by an animal to fellow animals or other species of animals to stimulate action(s) in the receiving animals [61, 62]. In other words, an animal vocalises to effect behavioural action(s) in the targeted receivers with the expectation that the receiving animals will appropriately use the acquired message from the vocalisation to take a responsive action [60, 63, 64]. For instance, the cries of an animal might be a message of attraction toward it or detraction away from it; and whatever action is to be taken by the animal(s) receiving the cries depends on the strength of pitches or frequencies of the vocalised sounds or cries [65] and in turn, the strength of the pitches or frequencies depend on the emerging stimulus in the environment of the animal(s) at a particular point in time. Consequently, vocalised sounds by animals at any point in time have distinct sound characteristics that denote the intent of the vocalising animal(s) and expected responses by the receiving animals. For instance, the croaking of male frogs described as 'whine' and 'chuck', maybe for attraction female frogs or keep other males away from the territory [63, 66]. Similarly, the barking or gnawing of a dog may imply a warning of a possible attack against and intruder. In the light of this, cries and sounds are essential tools by which animals communicate, not only with their surroundings [67] but also to convey a great deal of information over long distances for a specific purpose. Thus, most animals rely on sound communication for social relations, protection of themselves, survival and understanding of their environment [68].

Alongside the auditory communication is the visual dimension of the animals' communication whereby animals display signals that are visible to fellow animals. Such signals may be gestures, body postures, and colouration, movement or positioning of the body by the communicating animals [63]. Consequently, animals communicate non-verbally by expression of signals that could be described as acts or structures, to convey information to recipients to elicits a response in the recipients [69]. For instance, claw raising by hermit crab which implies an intention to attack is a warning communication signal to recipients and processing of such a signal would make the recipient respond by fleeing the environment of moving away from the crab. Similarly, a chimpanzee communicates threats by raising its arms, slapping the ground or staring directly at another chimpanzee [63]. Given the need for mating, some animals communicate to one another by a display of colouration change to attract the opposite sex. For instance, a display of bright yellow feathers by a male American goldfinch is a communicative signal to a prospective mate for mating. Animals thus communicate by behaviours to help them recognise and care for the young ones, find mates, coordinate group behaviour, defend territory and establish dominance.

Given the concept of animal communication, the same principle applies to farm animal communication to farmers whereby farm owners rely on behavioural actions of their animals as an element of information communication with them [66, 70–74]. In essence, farm animals communicate their current status to their owners by both vocalisation and behavioural signals. Given that environmental stimulus and/or hormone concentrations affecting the mood, thirst, hunger and appetitive behaviour of an animal stimulate specific behaviours and vocalisations in the animals, livestock farmers would need, not just to perceive the displayed actions by their animals, but to appropriately interpret such vocalisations and behaviours for necessary action to be taken. Although animals' vocalisations have been categorised into five 'main syllables' based on the mouth, tongue and nasal placement and the speed of air leaving the throat, no specific meaning has been attributed to different calls [75]. Consequently, a particular farmer will have to use his judgement, particularly based on experience and familiarity with his sets of animals, to determine the state of his animals' welfare and/or needs. Understanding the varying pitches or frequencies of sounds by animals is thus crucial to ensure an appropriate response to the animals' needs by a livestock attendant. For instance, a livestock attendant with a good understanding of the varying vocalisation pitches may get to know that a high and consistent pitch of sounds by a dam is an indication of out of contact with its kids [76–78] or that of difficult parturition. Frequency and pitch of coughing by farm animals equally serve as a good auditory cue to attract the attention of a farmer for the healthcare of the animal. Thus as an animal becomes more excited or distressed, the duration, volume, and pitch of the calls increase.

The auditory communicative signal may though be a narrow lead to early detection of illnesses in farm animals, the visual communication cue is more elaborate and much obvious to attract the attention of farm attendants or farmers for prompt veterinary actions. Ordinarily, a healthy animal is active and mostly in group association, but becomes dull or isolated when its normal healthy condition is impaired which becomes a communicative signal of ill health. As highlighted in **Table 1**, the displayed signals can be readily seen and obvious to stimulate veterinary actions.

Farm animals	General signs of illness	Respiratory signs	Gastrointestinal signs	Skin signs	Neurological signs
Cattle	Fever, lethargy (lack of energy) Excessive salivation Lameness	Multiple coughing animals Difficult laboured or rapid breathing Nasal discharge	Appetite loss Diarrhoea Abdominal pain Weight loss Dehydration (sunken eyes, prolonged skin tent)	Blisters or ulcers around muzzle, mouth, lips, gums, tongue, teats and/or feet Severe itching Circular areas of hair loss Warts	Behavioural changes or easily startled Restlessness or agitation Lack of coordination or high stepping Head rubbing, tossing or pressing Exaggerated blinking and chewing movements Trembling or convulsions
Sheep and goats	Depression Fever Lethargy (lack of energy) Dull coat Watery eyes Lameness, arthritis, or hot and painful feet	Coughing Nasal discharge Difficulty breathing	Weight loss Decreased appetite Diarrhoea and abdominal pain Food coming out of nose or mouth	Itchy, dry brittle fleece Red mouth and nose Blisters or ulcers around muzzle, mouth, lips, gums, tongue, teats and/or feet	Behavioural changes Excessive scratching and rubbing Loss of coordination Abnormal gaits (high stepping) Biting feet and limbs Head tilt or head pressing Inability to rise or convulsions
Pigs	Lethargy (lack of energy) Fever Lameness, painful movement and stiffness Swollen joints	Coughing Difficult laboured or rapid breathing	Lack of appetite Weight loss Diarrhoea	Blisters or ulcer around nose and feet Pustules Blotchy Skin	Behavioural changes Lack of coordination Excessive salivation or drooling Seizures or tremors Paddling while lying on their side
Poultry	Lethargy (lack of energy) Depression Drop in egg production Eye discharge Thin-shelled eggs Ruffled feathers Off feed or water	Open mouth breathing Nasal discharge Sneezing Coughing or gasping	Diarrhoea Weight loss Swollen crop or abdomen	Swollen and/or discoloured comb, wattles, legs and head Dark or crusty spots on comb Mites or Lice (near vent) Scaly legs	Difficulty walking Torticollis (twisted neck) Lack of coordination or inability to rise Tremors or paralysis of limbs or neck Paralysis and dilation of the crop Blindness Head pressing

Source: Developed from Department of Agriculture and Rural Development, State of Michigan Bulletin (Retrieved from online, July 1, 2019).

Table 1.
Signals for early detection of ill-health in farm animals.

Body movement of the animals is also essential cues to understanding the health communication behaviours of the animals. For instance, cattle will normally have their tails raised and positioned horizontally when defecating or urinating bur observation such positioning of tail aside the need for excretion is an indication of the health issue to be given attention. Also, kicking and tail swishing may be performed in response to acute pain with these signals directed toward the painful stimulus [49]. The ability of farmers to understand the communicative signals, however, depends on their good knowledge of different animal diseases and the signs that may be shown before the emergence of a particular disease. Based on experience, some farmers have developed the skills and intuition to rightly interpret behavioural signals of an animal about a specific kind of illness that is most likely to emerge or already discomforting the animals[2]. A field experience by which a set of goats' communicate their owner that their pen is heavily infested by lice was by their reluctance to enter the pen each day they return from free-range, and when forced into the pen, they began an unusual and constant stamping of their feet. But a curious examination of the sudden reluctance of entry into the pen revealed that the dusty floor of the pen was highly infested by lice which always walk into the goats' underneath hair thereby causing skin irritation to the animals. This was discovered when a multitude of lice flung onto the farmer's legs on entering pen thereby necessitating thorough cleaning of the pen and thereafter, no constant stamping of feet was observed among the goat. In essence, every behavioural actions farm animals might mean a lot and as such, livestock farmers need to understand and be able to distinguish between normal and abnormal behaviours of their farm animals to ensure good management of the animals' welfare. But farmers who do not have the experience or skill to accurately interpret the observed health communication behaviour of a farm animal would have to consult the service of a veterinarian. Thus farmers would have to consciously monitor the social actions of their animals for quick detection of laden diseases in their stock.

7. Essential healthcare services to attaining cost-effective health management of farm animals

Prevention and control of pests and diseases of farm animals are essential to achieving profitable and sustainable farm animal production. This involves putting up all necessary actions to ensure that animals in stock are free of infections or debilitating effects of pests and diseases. Such actions are not only to save the animals but also give an added opportunity of eliminating or reducing the cost of treatment, which is usually expensive where animals were to be treated. However, an important step to preventing and controlling disease outbreak in farm animals by farmers is cognition of health communication behaviours of the farm animals, nature and virulence pathogenic organisms that may induce ailment, and signs that are disease-laden and symptoms of emerging diseases in the animals. Because of this, farmers may have to promptly and diligently take the following actions for good health management[3] of farm animals and as well ensure profitable and sustainable livestock production.

[2] Not all diseases in farm animals could be determined by physical examination. It may require clinical examination, immune system function, nociceptor response and behavioural assessment for accurate diagnosis of a particular disease which of course could not be done a farmer but by a veterinarian.

[3] This section and of course this chapter did not provide information on treatment, prevention, and control of specific diseases but on basic actions that could be taken in health management bused on animal attendants' cognition of the animals' vocalisation and the observed visual cue for health management of the animals.

7.1 Surveillance

Effective care of farm animals begins with a vigil on the posture and environment of the animals. This entails regular checks on kept animals in stalls and fields and intermittent physical examination of the animals' bodies for early detection of impairment of their normal behaviours and any possible ill-health. Given this, farmers must be on the lookout for signs and conditions that may engender disease outbreak or infestation of pests in their animals. Hence, the understanding that a healthy animal will normally be on the stand on sighting their attendants and/or move excitedly when being approached to be fed makes it possible for a farmer to know that failure or refusal of an animal to stand up or move is a sign of impairment of the normal health of the animals and as such would need to be attended to, at least for examination and determination of what the health issue might be. Field experience in this regard with a livestock attendant On small ruminant farms shows that regular checks on the animals daily made it possible for quick detection of impairment of the animals' normal health condition. Observed cases of health issues in the stock as a result of regular checks on the animals include lameness (**Figure 4**) and refusal of one or two of the animals to move on being an approach. With these signs, cases of foot rot or scald, pneumonia, infestation by sheep fleas and worms in the animals. Physical examination of bodies of the animals for insects or pests is equally of great value in early detection of flies' infestation in ruminant skin and lice in poultry birds. Surveillance though begins with individual farmers and their farms, collective efforts become crucial to curtailing the spread of infectious diseases from farm to farm. Surveillance is about disease identification and reporting cases of infections by livestock keepers to animals' health agencies and veterinarians to enable disease patterns to be monitored [79]. On this note, the World Organisation of Animal Health—OIE, emphasises that effective surveillance system entails identification and/or observation of emerging disease and reporting such for sample and data collection, epidemiological and laboratory investigations, and management and communication of the resulting information to provide guidance on priorities and targets for the application of interventions to effectively control of the disease [80].

7.2 Hygiene practice

Hygiene practice is an important aspect of animals' health management and this entails keeping the farm environment free of any anything or condition that could induce pest infestation and disease infections in the animals. This includes

Figure 4.
A lamb with lame foot as result of infection. Source: Photo by author.

ensuring that farmhouses and animal beddings are dry and regularly cleaned, and as well as the equipment and all facilities used for movement and care of farm animals. Cleaning may include scraping, sweeping, washing, possibly with the use of disinfectants. This action is underscored by the understanding that pathogenic micro-organisms and eggs of external and internal parasites could survive under any condition, particularly in a wet and filthy environment. With hygiene practices, no favourable condition is created for the micro-organisms to be active or survive to cause infections in the animals. Disinfection may though be essential, most disinfectants are not particularly effective in combating viruses *thereby* emphasising the need for physical cleaning and burning of bedding [81]. Droppings or faeces by animals, feathers and dead animals need to be promptly removed pens and ranches as these are sources of pathogens [82]. Also, too many animals on-site and over-crowding within building and cages should be avoided to prevent the rapid spread of emerging diseases of stock [16].

7.3 Biosecurity checks

This is an aspect of farm hygiene but goes beyond the physical cleaning of farm facilities and equipment. The focus is basically on the cleanness of the animals concerning freedom from diseases of any kind. Scrutiny of animals to be introduced into farm sites is essential to maintaining farm hygiene and preventing the introduction of diseases into the farmyard. In the same vein, non-farm workers or visitors should be prevented from gaining access to the stock, be it in stalls or farm sites; and where necessary, all should have their feet and booths washed when going into the farm. Quarantining of new animals is essential to allow time for the manifestation of hidden disease in the animal(s) and such restriction may be for at least 3 weeks and/or possibly conduct a clinical test on the animal for the potential disease of concern. Other biosecurity measures to be taken include culturing milk from individual animals for contagious organisms, selection of healthy animals and use of semen, embryos or bulls from suppliers with control programs for the infectious disease [83].

7.4 Exploration of animals' behavioural communication cues

With a cognition of communication cues of farm animals, a keen observation of postures of the animals is crucial to ensuring quick response to the health needs of the animals. Farmers should know that a healthy animal is alert and aware of its surroundings by standing and actively holding its head up watching what is happening around it, be in close groups, moves easily and steadily with regular steps. Deviation from these postures implies impairments. Also, the eyes of the animals must be bright, ears must be erect and move swiftly in the direction of sounds and to get rid of flies, mouth, and nose must be free of dripping saliva and nasal discharge. Where discharges are observed in these body parts, it would imply health issues that need to be attended to. Examination of bodies of the animals should reflect smooth and shiny hair or coat if healthy, breathing should be normal, urine must be clear and faeces must not be watery [84]. Vigil on these cues is essential to initiate a quick response to the health needs of farm animals.

7.5 Disease diagnosis

While it may be possible to see visual communication cues of farm animals means of monitoring or detecting health issues in the animals, internally developed diseases may not be so easy until the animal reaches a critical stage of illness.

However, intermittent collection of animals' faeces for an examination of worm-eggs and examination of urine for blood and yellow colouration that could signify jaundice in the animals. The use of faecal egg counts has proven to be a valuable tool for detecting worm infestation in animals and the basis for designing appropriate deworming routine and determination of the right deworming medication. For instance, faecal analysis of cattle in Malawi reveals infestation of bacteria—coliforms and *Clostridium perfringens* spores; pathogenic protozoa—*Cryptosporidium* and *Giardia;* and enteric viruses—adenovirus, enterovirus, and reovirus in the farm animals. Examination of watery stool suggests diarrhoea in the animals and as such, necessary medication could be administered for treatment of the animals before their health situation result in economic losses [85]. Other diagnostic tests for detection of animals diseases include parasitological tests for detection of parasites in animals, microbiological and virological test for identification of the presence of micro-organisms and viruses in the animals, and serological or blood test for anlysis of blood serum of sampled farm animals. Other tests may include necropsies, abortions and stillborn, and milk tests.

7.6 Vaccination

This is an essential way to prevent or strengthen the immunity of farm animals to infectious diseases[4]. With the understanding that viral diseases cannot be treated vaccination becomes the means to prevent the outbreak of the diseases in farm animals. Vaccination protects the welfare of farm animals by preventing or reducing disease, which in turn reduces the pain and suffering often associated with illness [86]. Further insight into the value of vaccination shows that the drug mimics infections to provide immunity such that the animals could not be overreached by the disease but healthier [87]. However, the vaccine must be disease-specific and appropriately administered as recommended by the veterinarian and by checking the recommended dosage, dilution rate, route of administration and all precautions. Animals must be injected on the recommended parts of the animal's body and ensure that the injected sites are clean and dry.

7.7 Treatment of diseases

Treatment becomes essential and inevitable where farm animals have become infected by pests and diseases[5]. Unlike vaccination that is meant to prevent disease or infections by boosting the immunity of kept animals, treatment is meant to eradicate or halt the debilitating effects of diseases in farm animals (**Figure 5**). Treatment, however, takes different forms, depending on the nature of the ill-health affecting an animal(s). It could be by prophylaxis, intravenous injection, dipping, isolation or culling [88, 89]. For instance, treatment of bacterial diseases such as Salmonellosis could be by the use of antibiotics such as ampicillin, chemotherapeutics and fluid therapy, isolation and general nursing [89] while worm such as liver-fluke is treated with medication such as oxyclozanide, nitroxynil,

[4] Vaccination, especially when it involves an injection, is delicate medical care and as such could not be undertaken by just anybody but by well-trained animal health attendants or veterinary doctors. Vaccination is most cases are based on the pre-knowledge of certain diseases in certain farm animals and as such consultation of well-trained animal health officers for an appropriate recommendation of vaccines and dosage is crucial good health management of farm animals.

[5] Treatment of animal diseases is generalised in this section and as such, farm animal attendants will need to know that specific disease condition requires specific treatment and will be subject to appropriate diagnosis by well-trained animal health officers or veterinary doctors.

Figure 5.
A lamb being treated of infection by an experienced animal health attendant. Source: Photo by author.

albendazole [88], etc. In addition, treatment of bloat disease—gas build up in rumen, requires the service of a veterinary surgeon who might need to insert stomach tube into the rumen to have the built-up gas released and in extreme cases, the rumen will have to be punctured on the left flank with surgical apparatus such as trocar and cannula [16]. This suggests that a farmer cannot handle all treatment of emerging diseases in their farm animals.

8. Social and economic implications of health management of farm animals

The goal of livestock management is to ensure the efficient production of animals and animal products for social and economic gains at all strata of the human social system. Achieving this goal implies that farm animals must be kept healthy at all times for enhanced productivity. An important way to maintaining good health of farm animals is by eradication of epidemiological diseases or reduce their debilitating effects to barest level through combination of hygiene practices, biosecurity and vaccination of animals as failure in this regard may lead to immediate loss of livelihood to those in the livestock sector, disruption of domestic trade or the cessation of access to international markets, and threats to public health [1, 8, 10, 83]. Hence, the need for mitigation of the impact of diseases of farm animals. Prevention and control of animal diseases not only prevent the loss of animals to death but largely reduce the burden of the debilitating effects of disease and associated suffering of the animals such that they can enjoy better health and welfare. In addition, consumers could then have a supply of safe and affordable food [85]. In the same vein, vaccination of farm animals greatly protect animal and public health, reduce animal suffering, enable efficient production of food of animal source to feed the burgeoning human population, and greatly reduce the need for antibiotics to treat food and companion animals [90]. This, however, requires responsible use of vaccines and antibiotics to prevent, not just to control a disease outbreak in the farm animals but to ensure the safety of products from vaccinated or treated animals for safe consumption by humans. Control and prevention of diseases are however with cost implications whereby vaccination and treatment of animals add to cost the cost of livestock production and to the national veterinary budget [85]. This notwithstanding, advantages of boosting farm animals' immunity through vaccination and recovery of animals from ill-health by treatment worth the efforts and cost than losing the entire stock by death as the regained productivity of the farm animals could help recover the expended cost on vaccination and treatment in the long run.

DOI: http://dx.doi.org/10.5772/intechopen.89738

Author details

Okanlade Adesokan Lawal-Adebowale
Department of Agricultural Extension and Rural Development, Federal University of Agriculture, Abeokuta, Ogun State, Nigeria

*Address all correspondence to: deoaks@gmail.com

IntechOpen

© 2020 The Author(s). Licensee IntechOpen. This chapter is distributed under the terms of the Creative Commons Attribution License (http://creativecommons.org/licenses/by/3.0), which permits unrestricted use, distribution, and reproduction in any medium, provided the original work is properly cited.

References

[1] FAO. FAO's role in animal production. Animal Production. 2017. Available from: http://www.fao.org/animal-production/en/

[2] Murphy SP, Allen LH. Nutritional importance of animal source foods. The Journal of Nutrition. 2003;**133**(11):3932S-3935S. DOI: 10.1093/jn/133.11.3932S

[3] Zhang Z, Goldsmith PD, Winter-Nelson A. The importance of animal source foods for nutrient sufficiency in the developing world: The Zambia scenario. Food and Nutrition Bulletin. 2016;**37**(3):303-316. DOI: 10.1177/0379572116647823

[4] Grillenberger M, Neumann CG, Murphy SP, Bwibo NO, Weiss RE, Jiang L, et al. Intake of micronutrients high in animal-source foods is associated with better growth in rural Kenyan school children. British Journal Nutrition. 2006;**95**(2):379-390. Available from: https://www.ncbi.nlm.nih.gov/pubmed/16469157

[5] Muslimatun S, Wiradnyani LA. Dietary diversity, animal source food consumption and linear growth among children aged 1-5 years in Bandung, Indonesia: A longitudinal observational study. British Journal of Nutrition. 2016:S27-S35. DOI: 10.1017/S0007114515005395

[6] Kaimila Y, Divala O, Agapova SE, Stephenson KB, Thakwalakwa C, Trehan I, et al. Consumption of animal-source protein is associated with improved height-for-age *z* scores in rural malawian children aged 12–36 months. Nutrients. 2019;**11**(2):1-20. DOI: 10.3390/nu11020480

[7] Fitzhugh HA, Ehui SK, Lahlou-Kasai A. Research strategies for development of animal agriculture. In: Branckaert RDS, editor. World Animal Review. FAO; 1993. Available from: http://www.fao.org/3/U7600T/u7600T07.htm

[8] Dillivan K. Economic Benefits of the Livestock Industry. South Dakota State University, Extension Crops Business Management Field Specialist; 2014. Available from: https://www.agweb.com/article/economic_benefits_of_the_livestock_industry_naa_university_news_release/

[9] Liinamo A, Neeteson-van Nieuwenhoven A. The economic value of livestock production in the EU 2003. Farm Animal Industrial Platform. 2004. Available from: http://www.effab.info/uploads/2/3/1/3/23133976/03_economicvaluelivestockproductioneu.pdf

[10] Upton M. The role of livestock in economic development and poverty reduction. Pro-poor livestock policy initiative: Working Paper No. 10. FAO; 2004. Available from: https://assets.publishing.service.gov.uk/media/57a08cace5274a31e000136e/PPLPIexecsumm_wp10.pdf

[11] Tikam K, Phatsara C, Mikled C, Vearasilp T, Phunphiphat W, Chobtang J, et al. Pangola grass as forage for ruminant animals: A review. Springerplus. 2013;**2**:604. DOI: 10.1186/2193-1801-2-604

[12] Tzanidakis N, Stefanakis A, Sotiraki S. Dairy sheep breeding. In: Low Input Breeds Technical Note. 2014. Available from: http://www.lowinputbreeds.org/fileadmin/documents_organicresearch/lowinputbreeds/tn-2-1-dairy-sheep-2014.pdf

[13] Countryside. Best dairy sheep breeds for a farm: Learn how to start dairy farming with sheep. 2014. Available from: https://

iamcountryside.com/sheep/best-dairy-sheep-breeds-for-a-farm/

[14] Animal Nutrition. Department of Animal Science, Texas A & M University; n.d. Available from: https://animalscience.tamu.edu/academics/nutrition/

[15] Pritchett JG, Thilmany DD, Johnson K. Animal disease economic impacts: A survey of literature and typology of research approaches. International Food and Agribusiness Management Review. 2005;**8**(1):23 Available from: https://ideas.repec.org/a/ags/ifaamr/8177.html

[16] Sainsbury D. Animal health. In: Soffe RJ, editor. The Agricultural Notebook. Oxford: Blackwell Publishing; 2003. pp. 566-598

[17] Turnton J. Causes of diseases in animals. 2000. Available from: https://www.nda.agric.za/docs/Infopaks/diseases.htm

[18] BAMN. An introduction to infectious disease control on farms (Biosecurity). 2001. Available from: https://www.aphis.usda.gov/animal_health/nahms/dairy/downloads/bamn/BAMN01_IntroBiosecurity.pdf

[19] GSFR. Pathogenic Organisms. 2019. Available from: https://globalfoodsafetyresource.com/pathogenic-organisms/

[20] Carroll KC, Hobden JA, Miller S, Morse SA, Mietzner TA, Detrick B, et al. Jawetz, Melnick, & Adelberg's Medical Microbiology. 27th ed. New York: McGraw-Hill Education; 2016

[21] Kaiser G. General characteristics of viruses. Biology Libre Texts. 2019. Available from: https://bio.libretexts.org/Bookshelves/Microbiology/Book%3A_Microbiology_(Kaiser)/Unit_4%3A_Eukaryotic_Microorganisms_and_Viruses/10%3A_Viruses/10.01%3A_General_Characteristics_of_Viruses

[22] Burrows W, Cornelius CE, Scarpelli DG, Burrows W. Animal disease—Non human. 2018. Available from: https://www.britannica.com/science/animal-disease

[23] Beck K. Characteristics of a bacteria cell. Sciencing. 2018. Available from: https://sciencing.com/characteristics-of-a-bacterial-cell-13714449.html

[24] Orr RM, Kirk JA. Animal physiology and nutrition. In: Soffe RJ, editor. The Agricultural Notebook. Oxford: Blackwell Publishing; 2003. pp. 371-428

[25] Romani L. Immunity to fungal infections. Nature Reviews Immunology. 2004;**4**:1-13

[26] Casadevall A. Accidental virulence, cryptic pathogenesis, martians, lost hosts, and the pathogenicity of environmental microbes. Eukaryotic Cell. 2007;**6**:2169-2174

[27] Refai MK, El-Naggar AL, El-Mokhtar NM. Monograph on fungal diseases of sheep & goats: A guide for postgraduate students in developing countries. 2017. Available from: https://www.academia.edu/32819397/Monograph_On_Fungal_Diseases_of_sheep_and_goats_A_guide_for_postgraduate_students_in_developing_countries_Colonnade_and_Flock_of_Sheep_Statues_between_the_first_and_second_pilon_in_Karnak_temple_123RF.com_Luxor

[28] Wilkinson A. Managing livestock skin diseases on-farm: Causes and treatments. Farmers Guardian. 2017. Available from: https://www.fginsight.com/news/news/managing-livestock-skin-diseases-on-farm-causes-and-treatments-41959

[29] Jacob J. Internal Parasites of Poultry. Extension Issues, Innovation,

Impact. 2015. Available from: https://articles.extension.org/pages/66279/internal-parasites-of-poultry

[30] Taylor M. Protozoal disease in cattle and sheep. BMJ Journal. 2000;**22**(10). DOI: 10.1136/inpract.22.10.604

[31] Sahinduran S. Protozoan diseases in farm ruminants. In: Perez-Marin CC, editor. A Bird's-Eye View of Veterinary Medicine. Rijeka: InTech; 2012. pp 473-500. Available from: http://www.intechopen.com/books/a-bird-s-eye-view-of-veterinary-medicine/protozoan-diseases-in-farmruminants

[32] Kaltungo BY, Musa IW. A review of some protozoan parasites causing infertility in farm animals. ISRN Tropical Medicine. 2013; Article ID 782609, 6p. DOI: 10.1155/2013/782609

[33] Martínez-Girón R, Esteban JG, Ribas A, Doganci L. Protozoa in respiratory pathology: A review. European Respiratory Journal. 2008;**32**:1354-1370. DOI: 10.1183/09031936.00022008

[34] Seyfarth RM, Cheney DL. Signalers and receivers in animal communication. Annual Review of Psychology. 2003;**54**:145-173. DOI: 10.1146/annurev.psych.54.101601.145121

[35] Corwin RM, Tubbs RC. Common Internal Parasites of Swine. University of Missouri extension bulletin; 2018. Available from: https://extension2.missouri.edu/g2430

[36] Pig Site. Endoparasites (internal parasite)—Background and history. 2019. Available from: https://thepigsite.com/disease-guide/endoparasites-internal-parasites-worms-nematodes

[37] Gadberry S, Pennington J, Powell J. Internal Parasites in Beef and Dairy Cattle. Division of Agriculture Cooperative Extension Service, University of Arkansas; 2011. Available from: https://articles.extension.org/pages/11022/internal-parasites-in-beef-and-dairy-cattle

[38] Jackson DE, Ratnieks FLW. Communication in ants. Current Biology. 2006;**16**(15):R570-R574. DOI: 10.1016/j.cub.2006.07.015

[39] Leger DW. Contextual sources of information and responses to animal communication signals. Psychological Bulletin. 1993;**113**(2):295-304. Available from: https://www.ncbi.nlm.nih.gov/pubmed/8451336

[40] Raven PH, Johnson GB, Mason KA, Losos JB, Singer SR. Animal communication. In: Biology. 10th ed. New York: McGraw-Hill; 2014. pp. 1144-1147

[41] Manteuffel G, Puppe B, Schön PC. Vocalization of farm animals as a measure of welfare. Applied Animal Behaviour Science. 2004;**88**(1-2):163-182. DOI: 10.1016/j.applanim.2004.02.012

[42] Meen GH, Schellekens MA, Slegers MHM, Leenders NLG, van Erp-van der Kooij E, Noldus LPJJ. Sound analysis in dairy cattle vocalisation as a potential welfare monitor. Computers and Electronics in Agriculture. 2015;**118**:111-115. DOI: 10.1016/j.compag.2015.08.028

[43] Murphy E, Nordquist RE, der Staay FJ. A review of behavioural methods to study emotion and mood in pigs, *Sus scrofa*. Applied Animal Behaviour Science. 2014;**159**:9-28. DOI: 10.1016/j.applanim.2014.08.002

[44] Pond RL, Darre MJ, Scheifele PM, Browning DG. Characterization of equine vocalization. Journal of Veterinary Behavior. 2010;**5**(1):7-12. DOI: 10.1016/j.jveb.2009.08.002

[45] Leliveld LMC, Düpjan S, Tuchscherer A, Puppe B. Vocal correlates

of emotional reactivity within and across contexts in domestic pigs (*Sus scrofa*). Physiology & Behavior. 2017;**1**:117-126. DOI: 10.1016/j.physbeh.2017.09.010

[46] Straw BE, Zimmerman JJ, D'Allaire S, Taylor DJ. Diseases of Swine. New Jersey: John Wiley & Sons; 2013

[47] Michel AO, Mathis A, Ryser-Degiorgis M. *Babesia spp*. in European wild ruminant species: Parasite diversity and risk factors for infection. Veterinary Research. 2014;**45**(65):1-11. DOI: 10.1186/1297-9716-45-65

[48] Mosqueda J, Olvera-Ramírez A, Aguilar-Tipacamú G, Cantó GJ. Current advances in detection and treatment of babesiosis. Current Medicinal Chemistry. 2012;**19**(10):1504-1518. DOI: 10.2174/092986712799828355

[49] Anonymous. Cattle Behaviour. n.d. Available from: http://www.publish.csiro.au/ebook/chapter/9781486301614r_Chapter4

[50] Scott P. Lameness control in sheep. NADIS Animal Health Skill. 2009. Available from: https://www.nadis.org.uk/disease-a-z/sheep/lameness-control-in-sheep/

[51] MLA. Nutritional: Animal health, welfare and biosecurity. Meat and Livestock Australia. n.d. Available from: https://www.mla.com.au/research-and-development/animal-health-welfare-and-biosecurity/diseases/nutritional/

[52] Farm Health. Pig Diseases. Duchy College; 2018. Available from: https://www.google.com/search?safe=strict&source=hp&ei=zTI_XazXFcLlgwfioqrwDA&q=protozoan+disease+

[53] Ashfaq M, Muhammad G, Shamsheer-ul-Haq, Razzaq A. Effects of livestock diseases on dairy production and incomes in District Faisalabad, Punjab, Pakistan. International Food Policy Research Institute—IFPRI Working Paper No, 023. 2014. Available from: http://ebrary.ifpri.org/cdm/ref/collection/p15738coll2/id/128595

[54] Mantip SE, Shamaki D, Farougou S. Peste des petits ruminants in Africa: Meta-analysis of the virus isolation in molecular epidemiology studies. Onderstepoort Journal of Veterinary Research. 2019;**86**(1):e1-e15. DOI: 10.4102/ojvr.v86i1.1677

[55] Balamurugan V, Hemadri D, Gajendragad MR, Singh RK, Rahman H. Diagnosis and control of peste des petits ruminants: A comprehensive review. Virus Disease. 2013;**25**(1):39-56. DOI: 10.1007/s13337-013-0188-2

[56] FAO. Economic analysis of animal diseases. FAO Animal Production and Health Guidelines. No. 18. Rome. 2016. Available from: http://www.fao.org/3/a-i5512e.pdf

[57] Dalloul RA, Lillehoj HS. Poultry coccidiosis: Recent advancements in control measures and vaccine development. Journal Expert Review of Vaccines. 2014;**5**(1):143-163. DOI: 10.1586/14760584.5.1.143

[58] Narrod C, Zinsstag J, Tiongco M. A one health framework for estimating the economic costs of zoonotic diseases on society. Ecohealth. 2012;**9**(2):150-162. DOI: 10.1007/s10393-012-0747-9

[59] WSPA. What's on your plate? The Hidden Costs of Industrial Animal Agriculture in Canada. World Society for Protection of Animals. 2012. Available from: https://d31j74p4lpxrfp.cloudfront.net/sites/default/files/ca_-_en_files/wspa_whatsonyourplate_fullreport.pdf

[60] Bradbury JW, Vehrencamp SL. Animal communication. 2019. Available from: https://www.britannica.com/science/animal-communication

[61] Villarroel A. Internal Parasites in Sheep and Goats. Extension Service: Oregon State University; 2013. Available from: https://catalog.extension.oregonstate.edu/sites/catalog/files/project/pdf/em9055.pdf

[62] Gillam E. An introduction to animal communication. Nature Education Knowledge Project. 2011;**3**(10):70. Available from: https://www.nature.com/scitable/knowledge/library/an-introduction-to-animal-communication-23648715

[63] Khan Academy. Animal communication. 2019. Available from: https://www.khanacademy.org/science/biology/behavioral-biology/animal-behavior/a/animal-communication

[64] Owen C. Forms of animal communication. 2019. Available from: https://animals.mom.me/forms-animal-communication-8127.html

[65] Sánchez C. Stress-induced vocalisation in adult animals. A valid model of anxiety? European Journal of Pharmacology. 2003;**463**(1-3):133-143. DOI: 10.1016/S0014 2999(03)01277-9

[66] Narins P, Feng AS. Hearing and sound communication in amphibians: Prologue and prognostication. In: Narins P, Feng AS, Fay RR, Popper AN, editors. Hearing and Sound Communication in Amphibians. Berlin: Springer; 2006. pp. 1-11. Available from: https://www.springer.com/gp/book/9780387325217

[67] Ankel-Simons F. Survey of living primates. In: Ankel-Simons F, editor. Anatomy of Primate: An Introduction. Academic Press; 2007. pp. 47-160. DOI: 10.1016/B978-012372576-9/50006-1

[68] Discovery of sound in the seas. How do marine animals use sound? University of Rhode Island and Inner Space Center; 2019. Available from: https://dosits.org/animals/use-of-sound/how-do-marine-animals-use-sound/

[69] Laidre ME, Johnstone RA.Animal signals. Current Biology. 2013;**23**(18):829-833. DOI: 10.1016/j.cub.2013.07.070

[70] Walker SF. Animal communication. In: Mn Mey JL, editor. Concise Encyclopedia of Pragmatics. Amsterdam: Elsevier; 1998. pp. 26-35. Available from: http://s-f-walker.org.uk/pubsebooks/pdfs/animal-communication98.pdf

[71] Brumm H. Introduction. In: Brumm H, editor. Animal Communication and Noise. Vol. 2. Springer; 2013. pp. 1-6. Available from: https://link.springer.com/content/pdf/10.1007%2F978-3-642-41494-7.pdf

[72] Landsberg GM, Denenberg S. Behavior problems of the senior cat. In: Rodan I, Heath S, editors. Feline Behavioral Health and Welfare: Prevention and Treatment. Elsevier; 2015. pp. 344-356. DOI: 10.1016/B978-1-4557-7401-2.00025-8

[73] Bee MA, Miller CT, editors. Preface. Psychological Mechanism in Animals Communication. Vol. 5. Springer; 2016. Available from: www.marina-koller.com/epub/download/id=1000858& type=fil

[74] Manser MB. Referents and semantics in animal vocalizations. In: Miller CT, editor. Bee, M. A. Heideberg: Psychological Mechanism in Animals Communication; 2016. pp. 223-250. Available from: https://www.zora.uzh.ch/id/eprint/134988/

[75] Scott PR. The challenges to improve farm animal welfare in the United Kingdom by reducing disease incidence with greater veterinary involvement on farm. Animals (Basel). 2013;**3**(3):629-646. DOI: 10.3390/ani3030629

[76] Wagner K, Barth K, Hillmann E, Palme R, Futschik A, Waiblinger S.

Mother rearing of dairy calves: Reactions to isolation and to confrontation with an unfamiliar conspecific in a new environment. Applied Animal Behaviour Science. 2013;**147**(1-2):43-54. DOI: 10.1016/j.applanim.2013.04.010

[77] de la Torre MP, Briefer EF, Reader T, McElligott AG. Acoustic analysis of cattle (*Bos taurus*) mother–offspring contact calls from a source–filter theory perspective. Applied Animal Behaviour Science. 2015;**163**:58-68. DOI: 10.1016/j.applanim.2014.11.017

[78] Seyfarth RM, Cheney DL. Meaning and emotion in animal vocalizations. Annals of the New York Academy of Sciences. 2003;**1000**:32-55. DOI: 10.1196/annals.1280.004

[79] Postnote. Livestock diseases. Parliamentry Office of Science and Technology: Houses of Parliament. 2011. Available from: http://www.ifst.org/sites/default/files/Livestock%20diseases%20POST-PN392.pdf

[80] OIE. Guidelines for Animal Disease Control. World Organisation for Animal Health; 2014. Available from: https://www.oie.int/fileadmin/Home/eng/Our_scientific_expertise/docs/pdf/A_Guidelines_for_Animal_Disease_Control_final.pdf

[81] Stuart BF, Bedell DM, Lindsay DS. Coccidiosis in swine: Effect of disinfectants on in vitro sporulation of Isospora suis oocysts. Veterinary Medicine, Small Animal Clinician: VM, SAC. 1981;**76**(8):1185-1186

[82] Sonaiya EB, Swan SEJ. Small-scale poultry production: A technical guide. In: FAO Animal Production and Health. Rome: Food and Agriculture Organisation; 2004

[83] Cooperative Extension. Livestock Biosecurity. 2016. Available from: https://articles.extension.org/pages/65877/livestock-biosecurity

[84] Pashu Sakhi Module. n.d. Available from: https://aajeevika.gov.in/sites/default/files/nrlp_repository/Pashu%20Sakhi%20Hand%20book.pdf

[85] Leahy E, Bronsvoort B, Gamble L, Gibson A, Kaponda H, Mayer D, et al. Proof of concept of faecal egg nematode counting as a practical means of veterinary engagement with planned livestock health management in a lower income country. Irish Veterinary Journal. 2017;**70**:16. DOI: 10.1186/s13620-017-0094-9

[86] RUMA. Responsible use of Vaccines and Vaccination in Farm Animal Production. 2006. Available from: https://www.ruma.org.uk/farm/responsible-use-vaccines-vaccination-farm-animal-production/

[87] Roth JA. Veterinary vaccines and their importance to animal health and public health. Procedia in Vaccinology. 2011;**5**:127-136. DOI: 10.1016/j.provac.2011.10.009

[88] Agriculture. Common Animal Diseases and their Management. Vikaspedia. n.d. Available from: http://vikaspedia.in/agriculture/livestock/general-management-practices-of-livestock/common-animal-diseases-and-their-prevention-and-treatments#section-19

[89] Exper System for Cattle and Buffalo. General Disease Prevention and Control Measures. n.d. Available from: http://agritech.tnau.ac.in/expert_system/cattlebuffalo/general%20disease%20prevention.html

[90] Axtell RC, Arends JJ. Ecology and management of arthropod pests of poultry. Annual Review of Entomology. 1990;**35**:101-126. Available from: https://pdfs.semanticscholar.org/a06c/1 39a3cf193df8119363eaa178c5192b00703.pdf?

Section 2

Livestock Farming and Nutrition

Chapter 4

Nutrition and Health-Management in Dairy Production

Albert Sundrum

Abstract

The 'barrel model' of an organism's resource allocation pattern represents the basics for feeding regimes in farm practice. Several objections can be raised against the underlying theoretical assumptions, the generalizations derived from them, and the application of the model in practice. The concept particularly neglects the role of glucose and the increased competition for it between lactocytes and immune cells. It also fails to recognize the large variation between and within dairy herds. Lack of success in reducing production diseases calls for a modified approach – one which not only deals appropriately with the large variation between and within the dairy herds but also strives to balance the existing productivity/animal health/financial trade-offs. Instead of following general procedures, nutrient supply and outflow via milk have to be adapted to the individual requirements and health risks. To do so, the percentage of dairy cows affected by production diseases and failing to cope is a key criterion. Benchmarking of production diseases could act as an orientation point for farmers to compare their own position to other farms and thus set realistic target figures. Furthermore, means and measures to achieve goals have to be validated in the context in which they are used.

Keywords: complexity, energy balance, glucose deficits, production diseases, target figures, role of animal science, reductionist approach

1. Introduction

Various field studies suggest that dairy farming in general has up to now failed to substantially reduce the prevalence of nutritional disorders and associated comorbidities [1–3]. According to LeBlanc [2], dairy production is challenged by the fact that 30–50% of dairy cows are affected by some form of metabolic or infectious disease around the time of calving. The knee-jerk reaction which insists we still do not know enough to considerably reduce the undesired side effects of production processes is not valid as long as the current knowledge is not adequately implemented. On the contrary, it could be argued that the seemingly never-ending search for further knowledge in the same direction can be blamed for preventing reflection and discussion about a possible need for fundamental changes in the strategic orientation of dairy farming. However, as long as research emphasis is placed on finding technical and genetic solutions for current problems, the impression is created, and will remain, that there is no need to consider modifying the actual dairy systems. Yet in light of the lack of success, the question arises whether the high prevalence of production diseases (Pds) is not in fact an inherent problem of the production processes across the dairy industry. In raising this basic question, it is not the intention of the following script to repeat or summarize the general recommendations found in literature and text books that claim to provide options for reducing these serious problems. Instead, the objective is to question predominant thinking patterns and to reflect on the weak points and driving forces that might be responsible for preventing effective progress in the reduction of Pds in dairy farming.

2. Nutritional disorders and production diseases

Production diseases are a negative side effect of the production processes. They are not merely a peripheral phenomenon, although often mistakenly dealt with as such, but are related, amongst other things, to the issues of animal welfare and food safety. Pds adversely affect productivity and reproduction and can have severe economic implications due to related failure and prevention costs [4]. Moreover, products of diseased animals are of inferior quality, a fact recognized by consumers, who are aware of the problems in animal production. Thus, the high prevalence of Pds gives rise to questions and discussions about production processes and what responsible management actually entails. Pds have been under discussion in animal science since the first Int. Conference on Production Diseases in Farm Animals in1968, an event which has occurred periodically ever since [5].

Disturbances of one or multiple metabolic processes related to the regulation of a particular metabolite in the body fluids are known as metabolic disorders [6] and are a manifestation of the cow's inability to cope with metabolic demands [7]. Clinical diseases closely related to a suboptimal nutritional management are, amongst other things, ketosis, milk fever, metritis, mastitis, and lameness [8]. The known interactions between various metabolic stressors, and their relationships to other diseases, particularly infectious and inflammatory diseases of early lactation, have become "a central focus of interest in the study of metabolic diseases in dairy cattle" [9]. Nutritional disorders and comorbidities have been comprehensively discussed elsewhere [10]. In the following, they are considered as production diseases.

Pds occur throughout the lifetime of dairy cattle but are never so pronounced than in the transition phase, the 6 to 8-week period centered on parturition, and which is known as the most challenging and critical period for a dairy cow during the lactation cycle [11]. Within this period, major physiological, nutritional, metabolic and immunological changes occur. The production cycle of the cow shifts from a non-lactating state to the onset of extensive milk synthesis [12, 13]. Cows have to adjust metabolically to the sudden increase in energy and nutrient requirements and supply. Gaps between nutrient demand and supply can coincidentally occur with substantial variations in the nutrient content of the diet and in the daily intake of dry matter (DMI). Dealing with this requires comprehensive adaptation and regulation of the metabolism. Desirable outcomes for farm management are: cows that are successful in adapting metabolically to challenges inside and outside of the organism with minimal to no disease events and a reduction in avoidable culling as well as cows with efficient productive and reproductive performances. Past intensive research conducted into nutritional requirements, physiological adaptation and metabolic associations with periparturient diseases of cows has not led to any substantial reduction in the prevalence of Pds. Despite the fact that solutions are still not clearly evident, most in the dairy industry continue to believe that there are tremendous opportunities to improve the health and reproductive performance of transition cows without compromising milk production [14].

3. Allocation of nutrients by the farm management

It is beyond dispute that allocation of nutrient resources to the farm animals by farm management is of high importance for the realization of both a high level of productivity and a low level of Pds. However, in farming practice the nutrient allocation is not always demand- and target-oriented. In general, the diet offered is either in the form of a total mixed ration ad libitum or in a combination of a feed mixture ad libitum, both supplemented with an assigned amount of concentrate via an electronic feeder. The allocated amounts of concentrate are deduced from a more or less accurately estimated level of milk performance of each individual dairy cow, while the total or partly mixed rations are generally formulated according to the average performance level of a herd or feeding group of dairy cows.

This approach, however, neglects the large variation in the requirements of the animals due to, amongst other things, inter and intra-individual variation in milk yield, body weight and, last but not least, variation in feed intake. Grouping strategy and feeding behavior as well as social rank between the animals have a considerable impact on the competition between them for space and feed, and thus on feed intake [15]. Accordingly, a large inter and intra-individual variation of feed intake is observed in farm practice [16]. The amount of daily nutrient and energy intake is a result of the interactions between the composition of the diet itself, the environment in which a diet is offered and various intrinsic processes [17]. On the other hand, the same dry matter intake (DMI) per cow and day can be achieved by altered frequencies and durations of eating time and meal sizes. Nutrient and energy intake can change dramatically in response to changes in diet composition or metabolic state. Feeding regimes on a farm might appear to be regular but hidden variations in the nutrient and energy supply can occur to a greater or lesser degree. Furthermore, the feeding rations offered can be quite variable in their composition, for example in the portion of roughage and concentrate, throughout the course of time, thus correspondingly in the availability for the animals within the digestive tract [18]. Furthermore, the proportion of single components can vary considerably due to imprecision in mixing and/or in allocation procedures.

In light of the numerous sources of variation, feeding rations offered in farm practice cannot precisely meet the requirements of an individual cow within a feeding group or herd. The gap between demand and supply underlies a considerable variation between the animals. In general, farm management lacks insight into the degree of the inter and intra-individual variations and discrepancies. Knowledge about the impacts of nutrient supply is often restricted to the outcomes of feeding regimes in terms of the individual milk yields and content on a monthly, seldom on a daily, basis. Many farmers have knowledge about the composition and ingredients of the feeding ration and they can base estimations about the required feed intake on the analyzed portions of the diet [19]. However, these estimated equations are only valid for one virtual cow but represent the average of a feeding group. Considering that the interactions between the numerous influencing factors, of which only few have been mentioned, create a virtually unlimited number and variety of combinations (even within one single cow, let alone a herd), the discrepancies between demand and supply can only be poorly predicted by traditional models of feed intake regulation [20].

4. Resource allocation within dairy cows

Available energy is used by animals during biological processes (chemical, active transport, mechanical, electrical and thermal work) which are essential for building, sustaining and enhancing biological structures [21]. To grasp the complex processes within the organism, Weiner [22] proposed the 'barrel model' of an organism's resource allocation pattern, defined as the partitioning of available energy and substrates into various essential life processes, and into body structures and tissues [23]. According to Rauw [24], "input constraints (foraging, digestion and absorption) are engaged in series, whereas outputs (maintenance, growth and production) are parallel and independently controlled. If the sum of the output rates does not match the input, the balance is buffered by the storage capacity of the system" (**Figure 1**).

Feeding regimes in farm practices are generally based on this model when trying to meet the estimated requirements of cows with an adequate nutrient and energy supply, and to assess the amount of milk that can be expected from the ingredients offered by the diet. The 'barrel model' seems to be quite plausible in explaining the balance between input and output variables, and in offering options for farm management to react to increasing demands in the course of increasing output of energy via milk by inducing an increase in feed intake, digestibility and absorption of nutrient resources, and thus an increase in the availability of energy for the intermediate metabolic processes. However, when viewing this approach from different angles, several objections can be raised against its underlying theoretical assumptions, the generalizations derived from them and the application of the model in farm practice. The objections relate to the issues of self-maintenance,

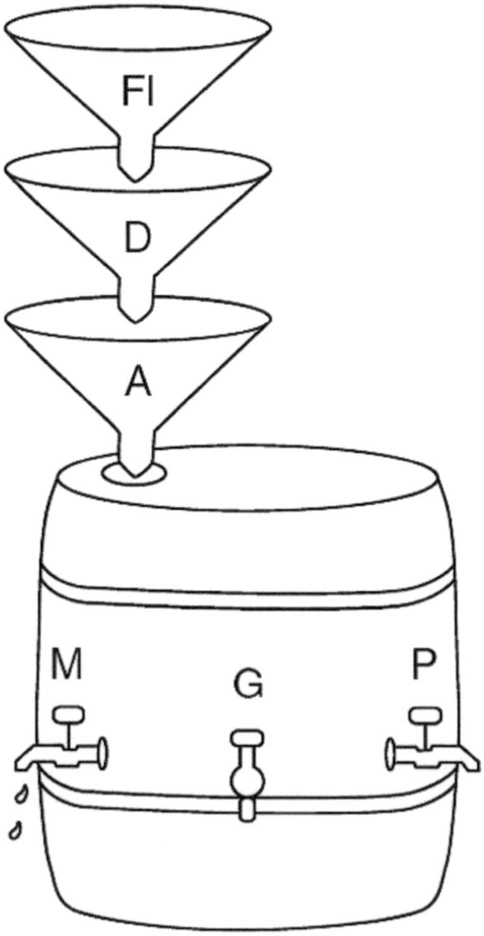

Figure 1.
The 'barrel model' of an organism's energy balance. The first spigot always leaks basal metabolic rate. FI, feed intake; D, digestion; a, absorption; M, maintenance; G, growth; P, (re)production [24].

storage capacity, regulation of allocation, intra and inter-individual variation, lack of evidence and the impacts on animal health and welfare. These issues are discussed below.

5. Available resources for (self-) maintenance

Not only a producing but also a non-producing animal expends nutrients to maintain its life processes. Every animal assures these expenditures have been met before allocating ingested nutrients to any other use. When comparing species of differing size and body weight (BW), it is known that smaller animals have a higher demand for maintenance energy than larger animals. The percentage ratio of regression to BW is approximately 0.75 and this is the accepted common base for the expression of maintenance requirements for nutrients across species [25]. Theoretically, it is an easy approach for calculating energy requirements of farm animals. It requires only reliable information on BW. Given the large variation in BW between dairy cows within a herd, and due to substantial changes which occur during the lactation period [26, 27], gaining reliable BW figures would demand a regular recording of individual BWs. This however is seldom consistently carried out in dairy farming.

Furthermore, cows of similar BW, size and breed may vary considerably in their requirements for basal metabolism. For example, locomotion activity is often considered as part of the 'maintenance requirement' even though the extent of activity, e.g. foraging, can differ extensively between cows. This also applies to the requirements needed to maintain the body core temperature because metabolically busy animals do not need additional energy to maintain body core temperature at the same ambient temperature as metabolically less busy animals. Much more significant is the fact that meeting the energy requirements for the basal metabolic rate and its variation due to differences in BW, activity level or cold and heat stress does not cover the requirements that are needed to ensure self-maintenance of the animal when faced with the various threats they have to cope with. McEwen [28] coined the term "allostatic load" within the concept of allostasis to describe potential permanent overburdening of homeostatic processes. One can imagine allostatic load increasing due to the rising energy expenditure required to fuel regulatory processes. Accordingly, allostatic load is the sum of the energy required to maintain basic homeostasis and to acclimate to changing environmental conditions.

The immune system, as one of the body's sensory organs for controlling interactions with the environment, is integrated into the physiological regulatory mechanisms that maintain the integrity of the host in the face of diverse environmental threats. Immune responses are not only influenced by the nature of the pathogen but also by characteristics of the host: age, gender, passive immunity, prior exposure to the pathogen, concurrent infections, physiological status, micro and macro-nutrient status as well as the presence of concurrent stressors [29]. The author identified demands on the immune system related to, amongst other things: "increased metabolic activity - systemically during fever, locally during activation of immune system cells; reduced nutrient availability due to anorexia and/ or other sickness; altered priorities for nutrient utilization due to changes in the gradient during immune activation that reduce the capacity of many non-immune tissues to utilize nutrients".

Whatever their origin, e.g. the accumulation of pro-inflammatory processes in dairy cows around parturition, disorders and diseases implicate the need of energy and substrates. These are needed to adequately meet the requirements of the immune response to prevent severe health problems of dairy cows and to support the overall goal of self-maintenance. According to Aitken et al. [30], "many aspects of the bovine immune system are compromised around the time of calving, especially the inflammatory responses". Immune suppression in the periparturient dairy cow is a commonly observed phenomenon and has been linked to poor metabolic status and negative energy balance [13, 31, 32]. Thus, attenuation of the immune response to the various challenges is, in the first place, the result of limitations in the availability of energy and substrates.

Besides energy, the organism requires protein, vitamins, minerals and trace elements in appropriate amounts in relation to the individual needs. Here the role of glucose must be emphasized: this substrate is used by the highly energy-demanding immune cells as a main source of fuel for immune defense and should therefore be considered the quantitatively most important fuel to fulfill the energy requirements of immune cells [33]. According to Ingvartsen and Moyes [34], "as glucose is the preferred fuel for immune cells, its low concentration during the transition period may partly explain the naturally occurring immunosuppression at this time". Another study [35] shows that ketones are not utilized by immune cells and in fact primarily act as inhibitors to immune responses when concentration is relatively high. In comparison with healthy controls, ketotic cows have increased circulating LPS prior to calving, and post-partum acute phase proteins such as LPS-binding protein, serum amyloid A and haptoglobin are also increased [35]. Endotoxin stimulates the immune system and activated leukocytes switch their metabolism away from oxidative phosphorylation to rely more on aerobic glycolysis [36]. The energetic cost of immune-activation is substantial but the ubiquitous nature of the immune system makes precise quantifying of the energetic demand difficult. Kvidera et al. [37] estimated approximately 1 kg of glucose is used by the immune system during a 12-hour period in which lactating dairy cows were challenged by LPS. This amounts to 2 kg of glucose per day for the immune system alone. This is more or less equivalent to the amount of glucose a cow withdraws in the form of lactose when producing an amount of 40 kg of milk per day [38]. In this context, it has to be considered that the synthesis of lactose alone utilizes 65–70% of the cow's total glucose turnover [39].

An increased immune system glucose utilization occurs simultaneously with infection induced decreased feed intake. This coupling of enhanced nutrient requirements with hypophagia further decreases the amount of nutrients available for the synthesis of milk. In the face of limited availability, glucose is allocated preferentially to a selected function at the expense of other functions, resulting in trade-offs and increased competitive pressure. Immune responses are context-specific and the costs vary considerably depending on the pathogen, the environment and defense capacity of the host. Immune defense activities create highly individual outcomes depending on the initial and boundary conditions, and on the degree of the mismatch between demand and supply. The context-specific and individual nature of immune activation suggests that quantitative estimates of the costs of immune activation can be neither readily generalized nor predicted. This might explain why the costs of immune defense are not a prioritized issue of animal science. Apart from striving for an easy approach to asses energy requirements of farm animals, it is quite astonishing and disturbing to realize that the costs of self-maintenance for farm animals are largely disregarded by the scientific discipline of animal nutrition. This may be due to the fact that feeding trials are generally conducted under standardized experimental conditions and usually on healthy animals which probably do not require high levels of additional energy and other substrates for the immune defense. Therefore, the results of feeding trials under experimental conditions enclose a high degree of uncertainty when transferred into practice on dairy farms where production diseases are frequently found.

The 'barrel model', with its focus on energy demand and supply, considers neither the requirement of glucose for adaptation processes nor the need for a glucose buffer to fuel a short-term activation of the immune defense. Furthermore, it is obvious that the allocation of nutrients and energy resources for the various needs is not regulated independently, as assumed by the 'barrel model', but is highly interconnected. In the past, it may have been rational for the diet formulation to consider maintenance, growth and production as separate singular outputs. In face of increased knowledge, it is not justified to maintain these assumptions any longer.

6. Storage of energy and glucose resources

Once nutrients are absorbed, they belong to the total energy and substrate pool and the stores of the body which make up the overall pool of resources. Metabolites in the form of carbohydrates, protein and fatty acids which are not needed in current metabolic processes are transformed and stored in the adipose tissues of the organism to serve as a reserve in periods of deficiencies. Fatty acids in the adipose tissues represent the main source of energy which the organism can fall back on when the supply does not meet the current needs. The 'barrel model' suggests that the energy balance is in general buffered by the storage capacity should the sum of the output rates not match the input [24]. Whenever intake is insufficient to support production – as in early lactation – energy is mobilized from body fat via the liver causing cows to milk off their backs and lose condition.

According to Friggens et al. [40], "high-yielding dairy cows have been genetically selected to partition even more glucose into milk production with the effect that reliance on body reserves has dramatically increased". Patton et al. [41] stated: "Even in the case of higher dietary intake, the increased input will primarily result in greater milk production while having little effect on energy imbalance and no beneficial effects on body condition and reserves at all". Although the storage capacities in the adipose tissue may seem to be abundant, the mobilization of fat can cause problems if the lipolysis is accelerated too fast. Metabolic changes, e.g. the process of uncontrolled lipid mobilization in response to excessive negative energy balance, increase the risk of ketosis, hepatic lipidosis and infectious diseases [42, 43]. Sordillo and Raphael [13] addressed the possible connections between fat mobilization and dysfunctional inflammation responses that may contribute to increased morbidity and mortality in the transition phase. Failing to adapt physiologically to an increase in nutrient requirements needed for the onset of milk synthesis is equivalent to metabolic stress and a major underlying factor in the development of transition cow disorders [44]. The authors conclude: "The combined effects of altered nutrient metabolism, dysfunctional inflammatory responses and oxidative stress can form destructive feedback loops that exacerbate metabolic stress and cause health disorders."

As mentioned above, dairy cows have a high demand for glucose. The storage form of glucose is glycogen and although widely distributed throughout the mammalian body, quantitatively the liver and muscle account for most of the body's glycogen stores. The liver glycogen depot plays a central role in intermediary metabolism, by storing and mobilizing glycogen during the metabolic states, with these responses modulated during pregnancy, lactation and exercise [45]. Metabolizable energy intake is the key driver. The glycogen depot in the muscle on the other hand is particularly important for local energy homeostasis. Compared with simple-stomached species, the rate of glycogen synthesis within ruminants is relatively low. Because ingested carbohydrates are efficiently fermented to short-chain fatty acids in the rumen, ruminants are required to meet the largest part of their glucose demand by de novo genesis [46]. A de novo generation of glucose by gluconeogenesis from non-carbohydrate precursors (e.g., lactate, glycerol, and amino acids) supplements the exogenous supply of glucose. Propionate is by far the predominant substrate for gluconeogenesis in ruminants [47]. The authors state that "the quantitatively most important adaption of metabolism to support the increased glucose demand in the immediate postpartum period is endogenous recycling of glucogenic carbon through lactate. This is mediated by a dual site of adaptation of metabolism in the liver and in the peripheral tissues, where the liver affinity for L-lactate is increased and glucose metabolism in peripheral tissues is shifted towards L-lactate formation over complete oxidation". Furthermore, the amino acid alanine is likely to contribute to liver release of glucose. If these adaptations fail, lipid metabolism may be altered. Increasing feed intake and provision of glucogenic precursors from the diet are important to ameliorate these disturbances. This applies in particular for an efficient gluconeogenesis because it is the major pathway for maintaining an adequate glucose supply. Glucose is, however, not only dedicated to the lactocytes in the udder, as emphasized by many animal scientists [39, 46], but is also as an essential fuel for many other cells and tissues of the organism. Thus, glucose needs to be permanently available at a sufficient level in the blood stream, and at the disposal of all cells which depend on it for their unimpaired operability. According to Bell [12], "daily requirements of glucose, amino acids, fatty acids and calcium for an early lactation cow are, respectively, more than 2.7, 2.0, 4.5 and 6.8 times greater than those needed for pregnancy. These differences represent changes in nutrient requirements over a short period of only one to two weeks, highlighting the tremendous metabolic alterations necessary to adequately support lactation." An imbalance in the glucose supply of high yielding cows in early lactation is unavoidable. The intensity of the imbalance is influenced not only by the level of milk yield and the degree of endogenous glucose provision but also by the demand of other essential tissues, inflammatory responses and, last but not least, by the immune defense.

In cases where dairy cows fail to cope with their living conditions due to the exceeding demands on their adaptation capacities, it is obvious that this is probably not only due to a lack of energy but also in particular to a lack of glucose. The concept of energy balance as represented in the 'barrel model' seriously neglects the role of glucose and especially the increased competition for it between the immune cells and the epithelial cells in the mammary gland. Furthermore, the 'barrel model' fails to consider that unpredictable events, including many biotic and abiotic stressors, need to be dealt with through immediate physiological and behavioral adjustments which can lead to situations in which the availabilities and the promptness of mobilization are overstressed.

7. Regulation of resource allocation

Biological regulation of the glucose balance within the organism involves a series of orchestrated changes; increased hepatic rates of gluconeogenesis, decreased glucose uptake and use by adipose tissue and muscle, a shift in whole-body nutrient oxidation so that less glucose is available as an energy source. First and foremost, the mobilization of essential resources requires well-functioning regulatory capacities to enable an efficient exploitation of resources to orchestrate a release of nutrients matching the requirements to a high degree and to deal with possible bottlenecks in metabolic pathways.

According to Baumgard et al. [39], the priority objective of the regulation is to ensure an adequate glucose supply to support lactation. A cow in a state of negative energy balance is considered "metabolically flexible" because she can depend upon alternative fuels (NEFA and ketones) to save glucose. In high yielding cows, the utilization of body energy reserves and the mobilization of body fat in the first month postpartum can be energetically equal to over one-third of the milk produced [48]. From a different perspective, the objective of regulation is to continually adjust the milieu to promote survival. Sterling and Eyer [49] introduced the term "allostasis" to refer to "changing regulatory systems ("stability through change"). Allostasis can be considered as the process of maximizing fitness in the face of environmental change and other unpredictable challenges. Regulatory mechanisms must change in order to maintain or achieve a state appropriate for the time of day or year and also in response to disturbances." From the perspective of the dairy cow, milk secretion is accompanied by substantial losses of energy and nutrients, particularly glucose from the body pool. A marked increase of cell differentiation and tissue hypertrophy in the udder is the starting point of an increase in milk yield [50]. According to Stefanon et al. [51], "the number of vital mammary epithelial cells control the initial conditions for the amount of milk produced as well as the amount of glucose needed for the production and secretion of lactose." Cows with a high genetic performance capacity for milk production are characterized by the ability to perform intensive gluconeogenesis and partitioning of the glucose into the udder while its contribution as a fuel source to extra-mammary tissues is decreased [12]. Because the uptake of glucose by the epithelial cells in the mammary gland is not insulin dependent, the cells have priority access to the glucose in the blood stream. According to Bauman et al. [52], "the productivity of this biological factory is extensive and in terms of the use of nutrients and energy, the cow should be viewed as an "appendage to the mammary gland" rather than vice versa."

Due to a sudden increase of nutrient requirements for milk production postpartum, a time when dry matter intake and nutrient supply lag behind, nearly every high yielding cow faces the challenge of shortages in energy and nutrients. According to Eastridge [53], "increases in genetic merit for milk yield go together with increases in feed intake but the latter does not fully compensate for the extra energy demands during early lactation. This results in a more or less extended negative energy balance and increased mobilization of body reserves." In order to sustain the various life-preserving functions, a limited availability of glucose provokes severe competition between different tissues in their need for glucose. It follows that "limitations require partitioning, and partitioning requires prioritization in guiding the nutrient flow to ensure that the demands of other cells, tissues and organs within the organism are not completely neglected" [48]. Accordingly, there is a need to avoid ruinous competition between sub-systems to prevent them from being swamped by unwanted side reactions which affect the viability of the system. Parasitic reactions by single organs at the expense of other organs may cause the whole organism to collapse. The question is, how and to what degree the regulation capacities are able to balance the trade-offs in their demands for glucose.

According to Lucy [54], "nutrient prioritization in early lactation to favor milk production over fertility is a reasonable strategy in biology. As nutrition becomes scarce, the lactating dam will preferentially invest the limited resources in the survival of living offspring rather than gambling on the oocyte that is yet to be ovulated, fertilized and cared for during an entire gestation. Selection for high milk yields takes advantage of the genetically programmed readiness of the dairy cow to enter into a negative energy balance at the onset of lactation and to mobilize resources from its body tissues." What is a natural biological process to ensure the maintenance of the offspring, however, might prove to be a self-harming trap when the selection process advances into dimensions that are far beyond the initial intention to ensure nutrient supply to the off-spring via milk. Dairy farming, and particularly breeding measures, takes advantage of the vulnerability of dairy cows in their self-defense against an excessive load by the demands of the mammary gland. However, milk production to safeguard the off-spring on the one hand and self-preservation of the dam on the other can come into life-threatening conflicts. This is the case when the gap between demand and supply gets to the stage where metabolic regulations are at risk of failing to balance the capacity of gluconeogenesis with the secretion of lactose, and of failing to mobilize the body resources needed to compensate for the deficits between nutrient output and intake. In general, the partitioning of resources within the organism is an excellent example of how cooperation works as long as there are enough resources available and as long as one part of the whole does not make unlimited demands at the expense of other parts. Shortcomings and problems can occur within several steps of the adaptation process, particularly those involving the adipose tissue and the liver. Further details have been explained elsewhere [10].

Generally, three options exist to alleviate the frequency and effects of these shortcomings and problems: (i) promote the absorption of resources to enhance availability or (ii) increase efficiency in the use of resources by partitioning the resources to those tissues and organs with the highest priority for the overall objective of self-maintenance or (iii) reduce the use of resources from the body pool as far as possible to sustain essential body functions. As absorption and partitioning have been optimized through a long-lasting evolutionary process, the major weak point lies in the limited capabilities to restrict nutrient losses via milk when it is necessary for the prevention of exhaustion due to overwhelming demands, and for self-maintenance. While the liver and muscle tissue have glycogen stores at their disposal, the mammary gland and the immune system rely completely on the body glucose pool. The body pool allows efficient trade-offs, that is, the organs grant each other short-term loans. If each organ were independently self-regulated, they would require their own reserve capacity, and thus more digestive capacity, to support an expensive infrastructure rarely used [55]. Efficiency in the use of limited resources requires organs to trade-off resources, that is, to grant each other short-term loans. However, milk secretion does not have an underlying central counter regulation which would enable a throttling of energy and nutrient losses via milk to prevent the dams from exhaustion and emaciation which subsequently weakens their adaptation capacities and risks their self-preservation.

The 'barrel model' illustrates that milking opens the flood gate for the loss of energy and nutrients, particularly glucose, from the body pool. However, energy and substrate losses, particularly glucose losses, can be so high that the minimum level required to sustain essential functions of the organism for self-maintenance is not maintained. The model is lacking possibilities to detect the filling state of essential resources in the body pool and thus those who are in charge lack information to throttle the outputs via milk to a degree necessary to leave enough resources for the processes of self-maintenance.

8. Dealing with inter- and intraindividual variation

The average milk yield per cow has increased considerably over the last decades, primarily as the result of genetic selection based on the moderate to high heritability of most production traits and the corresponding improvements in feeding regimes. Although animal breeders have accomplished a great deal in the past, they are not satisfied with these accomplishments. Accordingly, it is no surprise that the partitioning of energy within the organisms has also raised their interest. For animal breeders, options that emerge from the fields of genomics, proteomics, etc. to incorporate genetic differences between animals into nutritional models represent an area of exciting opportunity to improve nutrient partitioning and productive efficiency [39, 56]. According to Baumgard et al. [39], it is in fact only when the coordination of nutrient use is inadequate or an imbalance occurs that animal well-being and performance are compromised. However, in contrast to the underlying assumptions that inadequate coordination and imbalances in nutrient supply are exceptions, there is profound evidence suggesting that this is in fact the rule. The reasons are multi-layered and encompass the degree of supply, the processes in intermediary metabolism, the total requirements, and their coordination at the farm and animal level. The main reason, however, is inherent in the production process and lies in the large variation in the living conditions within and between dairy farms and in the intra and inter-individual variation at the animal level.

Dairy cows live under quite heterogeneous nutritional and environmental conditions and the individual animals themselves differ highly in their condition, their reaction and adaption capacities and, therefore, in their adaptive success. For example, while calculated energy balance is typically most negative within the first 12 days postpartum [57], differences amongst cows in time and extent of nadir and total energy deficits are large. In their study, the authors revealed that "over the course of 122 lactations mean values of total energy deficits during early lactation amounted to 1451 MJ NEL with a standard deviation of ±1062 MJ NEL. The postpartum interval to nadir of the estimated energy balance averaged 48 ± 29 days." Moreover, cows differ considerably with regard to the partitioning of energy and glucose between different physiological systems. Thus, cows with similar energy intakes and expenditures via the milk may actually experience differences in the burden of NEB and the shortage of glucose. This is not only based on genetic make-up (e.g. high v. low genetic merit) or the stage of lactation but varies greatly between individuals of the same genotype or in the same stage of lactation [58]. In their study, which allowed "discrimination between the roles of genotype (G), environment (E) (e.g. feed caloric density and milking frequency) and GxE interactions, the effects of genetic merit and milking frequency were significant only in the groups that were fed rations with high caloric density. However, signs of severe deficits in the availability of energy, poor protein balance and low body condition scores were not concentrated in the highest producing cows." Regardless of genotype, a reduced energy supply and extra milking had strong unfavorable effects on both energy and protein balance.

Large variations exist not only in terms of input and output but also in the availability of the various nutrients within the body pool. Substantial day to day variations in digestion and fermentation processes in dairy cows cause considerable variations in the relative quantities and supply of essential nutritional elements from the intermediate metabolic processes. This variation in the size of supply represents a real challenge for the metabolism in the face of demands for milk production and self-maintenance. Animals kept under highly standardized conditions on a research farm showed remarkable differences in changes in the concentrations of metabolites and hormones during the postpartum period [59, 60]. These findings indicate that the ability to cope with metabolic stress varies considerably between individual cows. On the other hand, energy partitioning between milk and body tissue can be altered considerably by diets that differ in lipogenic and glucogenic nutrient content [61, 62]. In addition, animals show enormous differences when confronted with various biotic and abiotic stressors and pathogens. In their reactions to changes in the environment, animals are not only influenced by the specific initial and boundary conditions, they also react self-referentially [63]. Unlike machines, their individual reactions cannot be predicted due to the interconnectedness of the numerous variables interacting with each other in a way that can switch from a more synergistic to an antagonistic relationship and vice versa.

In light of the large variation in biological processes and the deriving high level of complexity, the "barrel model" approach seems to be comparatively too simple and thus is not suited to be used in breeding and system biology to deal with differences between animals in the partitioning of nutrients. The approach lacks appropriate options to assess and deal with the intra and inter-individual variation of animals in their ability to cope with the highly variable internal and external challenges. The large amount of data harvested by "omics" techniques are noncausal. Nevertheless, representatives of "omics" research claim to demonstrate functionality and to develop a more comprehensive understanding of the regulation of the physiological processes and their role in animal productivity and animal health while applying the descriptive information gained from their research [39]. However, without accounting for either the large intra and inter-individual variation at the animal level or the variation in the living conditions at the farm level, it seems rather over-ambitious and presumptuous to claim accurate interpretation of the correlated changes.

A recent study [64], conducted on rabbits, provides some interesting observations and conclusions. Observing the resource allocation in different maternal rabbit lines revealed that the so-called "generalists" were able to appropriately allocate their resources to production, reproduction and health under suboptimal environmental conditions. The so-called "specialists" with high prolificacy were not able to allocate sufficient resources into health and reproduction. The authors concluded that the environment in which the animals are selected clearly drives the interplay between functions within the organism. If the objective of a selection program is to improve the overall fitness of animals without impairing productivity, a strategy is required that strives to establish a line of generalists.

9. Need for facts instead of assumptions

While the success of breeding programs in increasing the performance of dairy cows is obvious, their contribution towards improving the capacity of the animals to cope with unbalanced metabolic situations and challenges remains questionable. Breeding follows a single-sided approach that does not cover the multifactorial development of disorders and diseases, i.e., the approach does not consider why some animals are able to cope better than others. Nor does the approach take into account the context in which the animals are challenged. This applies not only to the respective conditions in which they live but also to the resources they can rely on or are lacking. In general, an external validation is not carried out because within breeding programs it cannot be assessed whether failures of animals in coping with the challenges are related to the genome or to the respective living conditions or to the interactions between both. Due to the lack of casual relationships, breeding programs can provide correlations but not explanations. Breeding programs do not demand, and cannot provide, a solution for the problems in the here and now. However, focusing and counting on breeding has the unbeatable advantage that it requires a fundamental change neither in the living conditions nor in the willingness to accept responsibility for the living conditions of the animals.

Primary causes and disturbing influences which contribute to the development of production diseases are manifold. They vary considerably between farms and animals. Some farms do well whilst others fail quite markedly in reducing clinical and subclinical problems, irrespective of average milk yields [11]. However, little is known about the causal network between the various factors involved in the uptake, partitioning and excretion of energy and nutrients [10]. Compounding the problem is the fact that variables such as feed intake, body condition, postpartum health and performance vary so widely amongst individual cows. Disturbances like change in diet, climatic conditions (heat stress), pathogen pressure, access reduction to trophic resources caused by competition with other individuals, injuries, diseases and other challenges can occur slowly or abruptly. The effects are disruptive and may be cumulative over hours or days or weeks. Additionally, factors such as an animal's current state of health and social status etc. may influence how it goes about its routines and how it responds to disturbances. Despite the highly heterogenous situation on dairy farms and the inter and intra-individual variation between dairy cows, the dairy industry still anticipates more robust and mechanistic models for predicting supplies and requirements of absorbed nutrients and available energy. Such models are expected to be useful in allowing for increased efficiency in the use of feed resources. Given the numerous influencing factors, the meaningfulness of models based on a few quantifiable variables is questionable, particularly when it comes to predicting the real outcomes. As adaptive success depends on the interactions between the level and type of threat, and on the current individual responsiveness of the cow, modeling this process is barely an option, let alone it providing information that allows for dependable prediction of outcomes. Nevertheless, modeling is suited to providing orientation towards possible outcomes (see explanations below).

High producing dairy cows are at a high risk of losing the capacity to cope with disadvantageous keeping and feeding conditions [58]. Diseases in animals are an indication that their physiological condition is out of balance [10]. This is often due to limitations and mismatches of resource allocation. Ingvartsen et al. [65] pointed out that evaluations found in literature on the relationship between performance and incidence of Pd are in all probability meaningless as inherent biological correlations – besides within and between-herd confounding effects – exist. According to Mulligan and Doherty [7], the hypothesis that high yielding cows automatically have higher levels of production diseases is likely to be as false as the hypothesis that lower yielding cows have lower levels of production diseases. Health problems are context-variable and need to be addressed in the context in which they emerge. Often, the degree of the clinical signs of disorders and diseases is neither assessed comprehensively nor monitored consistently, let alone always tracked down to the possible causes [66]. Achieving a low prevalence of production diseases is rarely considered as an independent production goal as it is easier to simply perceive them as being an unavoidable negative side effect of production processes. The multifactorial background of production diseases as the result of overstressed adaptation capacities hinders easy identification and solving of health problems. Commercial farms can seldom provide conditions that allow observations to be performed on a ceteris paribus basis as under experimental conditions. In contrast, impacts of the various influencing factors on the ability of farm animals to cope are not constant and do not emerge separately from one other. Adaptation is a functional and target-oriented process involving the whole organism and thus cannot be narrowed down to single factors [10].

Instead of following general assumptions and mental associations about possible relationships between single variables and the impacts of management measures on these variables, an obvious step when striving to solve health problems in the here and now is to estimate the degree of metabolic disorders and associated comorbidities at an individual farm level. Doing this requires regular monitoring, an indispensable component of any serious attempts to develop context-specific strategies regarding the improvements of production diseases. This alone, however, is not enough. It needs to be supplemented with the acquisition of further data on individual cows, particularly the degree to which the energy and nutrient supply correlates to the individual needs of an animal according to its specific stage of life and living situation. Yet, generally speaking, even when available, farm management is often not able to correctly interpret data regarding the negative energy balance of the individual cows and thus cannot know which animals are at a higher or a lower risk, which animals are able to cope with the NEB and which ones are showing disorders as a sign of adaption stress due to whatever reasons.

10. The role of farm management

Modern animal production is mainly based on economic principles; neither animal health and welfare nor ecological issues are taken much into consideration. However, the high morbidity and mortality in dairy production associated with poor welfare conditions necessarily questions modern industrial farming practices. In the past, politicians, agronomists and animal scientists have appeared to assume an ability in the markets and in science to drive technological changes which will enable the economic-environment system to both satisfy increasing global food demands and simultaneously solve problems of animal health and welfare. These key assumptions, however, have turned out to belong to the category of wishful thinking, lacking as they are in any profound evidence. They generally neglect, on different scales, the biological basics, particularly the complexity of physiological processes as well as the ambivalent nature of productivity and the resultant trade-offs. Whether processes are beneficial or non-beneficial very much depends on the context in which they take place and the level at which the situation in question is being analyzed. The same is also true for the possible options of balancing the trade-offs between economic interests and animal health and welfare in a cost-effective manner. Thus, the frequent attempts to formulate one-size-fits-all general recommendations for a successful implementation of measures are often misleading and contradict the actual context-specific nature of biological processes and the subsequent need for context-specific solutions at all levels of dairy farming.

10.1 Lack in orientation

For the farmers, it is often very important to know where they stand in relation to other farms. Data from a representative number of farms could be used to create a scale ranging from very low to very high prevalence of Pds per farm unit thus giving farm management an idea and orientation as to whether the individual farm belongs to the category of farms with a low, a middling or a high level of health problems. However, as long as data on production diseases is not sufficiently solid, it has little practical value for farmers and they can basically disregard it. Thus, a diagnostic procedure is essential for the assessment of the prevalence of Pds as well as for the identification and implementation of measures appropriate for the farm specific situation and for the need to balance partly contradicting goals. Both the rate of productivity and the prevalence of production diseases on dairy farms emerge from very complex processes. Focusing on single aspects without taking into account both the context and the conflicts between achieving productivity and the development of production diseases does not allow any truly valid statements and can be said to be overly narrow.

In contrast to zoonotic and epizootic diseases, productions diseases are not yet a matter of public concern. Although accompanied by pain, suffering, distress and longer persistent harm, they are not even fully recognized as a severe animal welfare issue. Being primarily treated as an internal affair of the farm, Pds are not regulated by legislation but left in the hands of the farmers and their individual readiness to act. Generally, personal economic concerns are a greater deciding factor than concern for the health of animals. Indeed, it is often said that farm management is interested in reducing Pds solely due to economic reasons. Indeed, there is no doubt that Pds can cause severe economic losses. However, there is still a lack in valid data about the degree of failure costs, and particularly on the effort in terms of labor time and investments required to reduce the prevalence of production diseases in the farm specific context [4]. To reduce production diseases in a cost-effective way and to maintain production at a competitive level is not only highly contextual and thus requires appropriate farm-specific measures but also relies on a function of margin utility. This applies to the use of resources e.g. high-quality feed, labor time, investments to reduce production diseases as well as to the intensification occurring in the increased use of inputs. Additionally, the conditions outside the system boundaries (in terms of the price of products sold, availability and price of resources needed to reduce production diseases) have to be considered. Cost–benefit relationships not only depend on the status quo for both productivity and production diseases but also on the gap between the status quo and the envisaged target figures.

10.2 Need for profound data

Optimization of the relationship between productivity and the prevalence of Pds to the benefit of both the farmer's income and the health and welfare of farm animals requires access to reliable farm-specific data. Thus, a major question is how to increase the availability of valid data and how to create an overview that can support decisions of management regarding an efficient allocation of available resources. Records of milk performance at an individual level – either through daily milk yield measurements or official milk recording – are valuable tools, and not only for performance monitoring. They also reflect the individual requirements of the dairy cows in the course of lactation and are thus essential for implementing a target-oriented nutrient supply. Often this data may be considered unnecessarily costly or time-consuming for flat rate concentrate feeding or TMR systems. Lactation curves plotted for individuals or groups of animals provide a very graphic illustration of performance. Since it is always one of the first things to be affected by the diet, milk components are an essential element for monitoring the impacts of energy and nutrient supply on dairy cows. Fat and protein levels are especially valuable indicators of diet adequacy. Furthermore, feed intake is a critical factor in providing the right degree of nutrition. Given the wide impact, dietary problems and imbalances can have on productivity, monitoring specific aspects of dairy health and fertility can be very valuable in feeding management. However, appropriate techniques for the assessment of feed intake at the individual level are not yet fully developed for use in farm practice. One sophisticated and cost-effective technology which is available but seldom implemented on dairy farms is measuring equipment which can continuously determine the body weight development. More often in use is the tool of Body Condition Scoring which provides information about the measure of a cow's energy balance. However, one disadvantage of this tool is that long temporal delays can occur in receiving the information on discrepancies and thus in subsequent responses to the information. The time delay between cows receiving too little energy in their feed rations and any resulting pregnancy rate problems often go unrecognized until it is too late to do much about them. This is compounded by the fact that fertility is not given sufficient prominence within rationing programs although fertility, like lameness, is now recognized as being caused by a number of factors, including inadequate nutrition. Furthermore, disorders like acidosis, ketosis, displaced abomasum and fatty liver are clear signs of dietary problems, especially if they affect a number of animals rather than just the odd individual and occur regularly rather than occasionally.

10.3 Dealing with complexity

Due to the fact that being confronted with the complexity of production diseases for a long time, farm practice is unswervingly in search of simple approaches to deal with unintended side-effects. This contradicts with the major challenge, namely variation, in dairy farming: variation in the energy and nutrient supply of the individual cow, variation in the utilization and partitioning of energy and nutrients and the differences in the requirements needed for the essential tasks of regulation and immune defense, resulting in a large variation in the gaps between supply and demand. Without feeding control measures, it is left to the animals to cope with the occurring discrepancies. In this respect, an apparent "survival of the fittest" selection occurs at the farm level. However, it is not necessarily a selection between farm animals based on comparable initial and boundary conditions but on a highly variable situation in terms of performance level and additional demands by fluctuating internal and external stressors. Consequently, the selection might occur to a high degree by pure chance.

The prevalence of production diseases on farms indicates the degree of shortcomings of farm management at two levels. Firstly, regarding the degree of a demand-oriented supply with energy and nutrients according to the individual requirements, and secondly, regarding protection against stressors such as pathogens, crowding effects or heat stress which lead to a need for additional resources. Production diseases are always context specific. The context is characterized by the specific farm conditions, the individual cow situation and the interactions between both. Dairy farms vary widely when it comes to the living conditions of the animals. Thus, health problems require a diagnostic procedure at the farm level. This diagnosis needs to include the most relevant influencing factors involved in the multifactorial processes as well as estimations about the most effective and efficient strategies in the farm-specific context [3, 67, 68].

10.4 Assuming responsibility

An important prerequisite for any improvements, however, is related to the need to assume responsibility not only for the results of efforts to increase productivity but also for the negative side effects of the production processes. The farm management designs the living conditions of farm animals and organizes the allocation of resources but it is also, to some degree, responsible for the resource partitioning processes within an organism. This applies to breeding for a high number of lactocytes in the mammary gland, which is responsible for a prioritized skim of glucose from the body pool, and to sucking away every last drop of milk via milking. Metabolic disorders indicate an imbalanced trade-off between the original goal of sustaining the offspring via milk and the goal of self-maintenance of the dam. Farmers are challenged to reduce the degree of trade-offs by adapting the breeding practices to the quality of available nutrients and, at least, temporarily, decreasing the amount of milk extracted during milking, perhaps also the frequency of milking. To increase milk production, it is not uncommon in intensive dairy systems to increase milking frequency to three times daily. Reducing milking frequency is much less common. In doing so, it is in fact possible to improve the overall energy balance of cows during early lactation with once-daily milking [69]. Furthermore, this procedure can entail an improvement in the metabolic profile [70] and immune function [71] of dairy cows. In contrast, Soberon et al. [72] reported that cows subjected to an increased milking frequency are 1.4 times more likely than the control cows to be classified as sub-clinically ketotic. Depending on the stage of lactation, breed, and parity, the reduction in milk yield losses in the course of short-term alterations to milking frequency in early lactation varies considerably and can amount up to 22% [73]. This figure has been revealed as an average milk loss across 30 different international short-term studies. While it is comprehensible that short-term alterations to milking frequency may provide a tool to better manage the metabolism and energy balance of cows during early lactation [74], many farmers fear that the losses in milk yield could be too high. All the more is it necessary not to go for a general strategy but to develop a farm specific strategy that suits the situation of the individual cows, thereby also considering the fears of farmers. The aim should be to throttle the withdrawal of milk by adapting the quantity to the estimation of risk for the individual cow. This practice would be suitable under certain infrastructural farm conditions, including the availability of valid data on dairy systems where an emphasis is placed on animal health rather than on milk production per cow.

10.5 Unfair competition

Finally, the crucial question is, at which rate of disturbances and Pds should an intervention by farm management take place. Currently this is determined solely by individual farm management, often to the detriment of the animals. Farm management, however, would be well-advised to show an interest in reducing the prevalence of production diseases. A high prevalence of production diseases represents a low health performance by farm management. Farms with a high prevalence of Pds are not only disregarding their obligation to prevent suffering of the animals but are also delivering inferior products to the market. Thus, low levels of Pds should be seen as a significant production goal which carry as much weight as productivity goals. However, setting low disease levels as a production goal will only occur when farmers realize that they can gain an advantage over competitors who have higher levels of Pds. On the other hand, farms behave unfairly when they cause, and/or basically ignore, a high level of Pds and related welfare problems and therefore produce an inferior level of product and process quality while simultaneously achieving the same market prices as those who invest time, money and effort in product and process improvements. Moreover, farm associations, like those of organic agriculture, should be more concerned about unsatisfactory health performances amongst member farms and doing more to raise the lack of concern shown by their consumer clientele as this defies general consumer expectations and any efforts to justify the premium prices [75]. Whether it is intrinsically motivated or forced by economic reasons or the demands of retailers to improve the current unsatisfactory situation regarding the prevalence of production diseases, farm management needs to know how and where to direct its efforts. Benchmarking would offer an appropriate methodological approach to deal with the issue of unfair competition and also with the uncertainties in the assessment of Pd data as these methodological uncertainties affect all farms, if not exactly to the same degree. Benchmarking allows target figures to be deduced from the average levels obtained from assessing a sufficient number of comparable farms or from an estimate of the optimum balance between productivity and disease related loss and failure costs. While farmers are generally hesitant in their readiness to extend control to others, farm management lacks orientation regarding its own position in relation to other farms and regarding the target figures it should aim for in the future as long as benchmarking is not established. The lack of benchmarking for Pd values in relation to the product quantity of products from animal origin can be seen as one of the main barriers in the fight to reduce nutritional disorders and related Pds in dairy production.

11. The role of animal science

Due to the lack of sustainable success in reducing production diseases, or at least the lack of evidence of a general improvement in animal health and welfare in dairy farming, the various disciplines of animal science are challenged to reflect on the possible reasons and on their own role in the overall process.

11.1 Predominant focus on performance

Increasing the milk performance of dairy cows is still the predominant goal in dairy production, primarily driven by economic considerations and supported by various disciplines of agricultural and animal science. Particularly the discipline of animal breeding continues to embrace this strategy, even though scientists make efforts to integrate functional traits in the breeding programs, so far with no truly convincing success. The traditional approach of animal breeding indirectly evokes the impression that it is the animals and/or the genome rather than the living conditions which are the real weak points in the system and, consequently, it is the former rather than the latter which needs to be further improved. It is obvious that a unilateral focus is not appropriate for dealing with problems that emerge from the interactions between various components within the organism, and between the organism and its respective living conditions. Dealing with the issue of animal health and welfare in dairy production cannot just be left to the predominant paradigms and interests of single disciplines. Instead, there is a need to first gain an overview and to identify the predominant weak points in the farm-specific context. A unilateral objective of increasing milk performance, together with a one-sided disciplinary focus are probably at the heart of the ignorance surrounding the negative side effects which accompany the production processes. On the other hand, what is being ignored cannot be solved.

In light of the farm-specific challenges, particularly the large inter and intra-individual variation of the gap between nutrient demand and supply, the scientific discipline of animal nutrition is not yet able to offer adequate tools to balance energy and nutrient input/output figures on an individual base. When the supply level is tailored to suit one virtual cow whose average values of nutrient and energy requirements act as the reference for a whole feeding group, then the variability in the requirements between the individual cows of a feeding group is widely disregarded. This also applies to the requirements an individual animal needs for regulation (allostatic load) and immune defense activities in relation to the supply of energy and nutrients, particularly glucose. The additional requirements are hard to predict and are thus blind spots in the discipline of animal nutrition. However, they cannot be disregarded any longer when the reduction of the prevalence of production diseases on the farm level is on the agenda.

While some scientific disciplines are engaged in furthering increased performance in animal production through, for example, breeding or feeding methods, other disciplines, e.g. veterinary science are trying to deal with the negative side effects of the intensification processes, also with no truly convincing success. The different disciplines seldom work together to find common strategies to deal with contradictory goals and the uncertainty regarding their effects. Instead, there is an enormous temptation for animal scientists to gain a scientific reputation by becoming a specialist who focuses on single areas at the risk of losing sight of the whole picture. This focusing by the numerous experts on their respective topics has led to a dissociation of the generalist approach. The re-integration of the subcomponents into a well-functioning whole requires an enforced interdisciplinary effort to focus on the performance of the whole system rather than on the separate optimization of individual components. It goes without saying that this is easier said than done.

11.2 Reorientation

In the future, it will not be sufficient for dairy production only to produce a high amount of milk in a cost-effective way. Milk production has to be carried out also in an animal and environmentally friendly manner, therewith considering the values of common goods. Realizing a comparable low prevalence of productions diseases is equivalent with the animal protection service of a farm system [76], which can be offered as a quality service on the food market. This approach allows the alignment of animal protection and production services of the farm system, appearing as a new production goal. The balance between animal protection and production services result from the entirety of processes which take place within a farm system. To succeed in this effort requires more than relying on a general scientific knowledge base, but needs a systemic, functional and result-oriented approach.

The traditional tools of improvements when dealing with complex issues are: problem analysis, defining a short-term and/or long-term goal, developing promising strategies, implementation of most appropriate measures, and finally adequate control and monitoring of success. This roughly sketched approach parallels with the deductive approach of veterinarians when examining, diagnosing and treating single diseased animals except that its focus is not on the recovery of individual animals alone but extends to the recovery of individual farms. Certainly, such an approach requires continuous acquisition of information on the nutritional status of the individual animals, the capacities of the living conditions and, last but not least, the resulting outcomes of the interactions between individual animals and their respective living conditions in terms of clinical and subclinical diseases. These indicate that animals are currently not able to cope. The percentage of dairy cows not being able to cope should be the key criterion for all subsequent activities.

11.3 Providing orientation and 'action knowledge'

Different kinds of internal regulations are required on the farm level. Animal science is asked to provide orientation and to develop 'action knowledge' to create strategies for management to sustainably allocate the relevant resources within the farm system, particularly considering the trade-offs in resource flows through various sub-systems. To formulate concise working hypotheses regarding the most effective and cost-efficient means that are at the farmer's disposal and to organize an appropriate allocation of resources within farm-specific contexts requires the determination of target figures in relation to the envisaged prevalence of Pds.

At the same time, the impacts of tools and means intended to reduce Pds have to be context assessed to establish whether they work effectively and whether they provide a positive cost–benefit ratio. Many technical tools and measures to reduce production diseases have been proven in scientific studies but nearly solely under standardized conditions. The results of these studies are at the farmer's disposal via mediation by advisory services and thus belong to the category of 'disposal knowledge' [77]. However, trying to find general solutions for the mitigation of negative side effects, e.g. in offering general recommendations in the field of breeding [39], technical developments or precision farming [78], might be blamed for oversimplification. By predominantly focusing on the development of 'disposal knowledge' in relation to single traits, animal science fails to grasp the complexity of the challenges at hand. This inductive approach distracts the focus from the problems occurring in the here and now and related to the farm-specific context. Simultaneously, it makes farmers believe that this might make the need to implement fundamental changes within the production processes seem unnecessary. 'Disposal knowledge' can claim to be valid only for the specific conditions under which it has been proven. When implemented in a specific farm context, it functions only as a working hypothesis for 'action knowledge'. The impacts that might occur in the use of generally recommended means and tools require external validation to assess whether they are able to contribute to the envisaged end and to deliver what they promise. This includes proving their suitability in contributing to alleviating the conflict between productivity and animal health on individual farms. Without external validation, general tools to reduce production diseases seem to be an end in themselves rather than a means to an end.

The extent of the outlined complexities explains why it is so difficult to improve the unsatisfactory situation regarding animal health and welfare in dairy farming. Too many partly diverging interests of different stakeholder groups, including the interests of animal scientists, are involved. However, if general enlightenment belongs to the crucial tasks of scientists, as they themselves maintain, this stakeholder group is under a particular obligation to consider animal health and welfare as belonging to the common good and are therefore obligated to contribute to improvements therein. However, as long as animal scientists claim to be able to offer simple solutions based on a reductionist approach without providing convincing evidence, they could be considered as part of the problem rather than part of the solution.

12. Conclusion

An oversized gap between nutrient requirement and nutrient supply in the transition period is a major cause for nutritional disorders and associated comorbidities in dairy cows. Due to the large inter and intra-individual variation in relation to feed intake, output of glucose via milk and metabolic adaptation capacities, the risk for the development of production diseases is a matter of the individual animal in the farm-specific context at a given time. Consequently, the risk cannot be predicted and solved by an inductive approach. The individual animal is the reference system for the appropriateness of nutrient and energy supply as well as the need for protection against biotic and abiotic stressors. Different animals need different supplies and a different kind and degree of protection. Correspondingly, the approach to deal with dairy cows that belong to a feeding group as a homogeneous unit and addressee of management measures is misleading.

Feeding regimes in farm practices are generally based on the concept of energy balance as represented in the 'barrel model'. However, it has become obvious that the allocation of energy resources for the various needs is not regulated independently, as assumed by the model, but is highly interconnected. But above all, the model neglects the role of glucose and the increased competition for it between the immune cells and the epithelial cells in the mammary gland. The model does not create a sufficient predictive power. Therefore, it is not justified to adhere to the model any longer.

The real challenge of nutrient management is to organize an efficient allocation of the available nutrient and labor resources closely related to the individual needs of the animals. Certainly, farm practice does not allow to feed dairy cows individually. However, there are various options - on the one farm more than on others - that should be taken into account when striving for a low level of Pds. Amongst other things, the implementation of feeding phases should be highly adapted to the corresponding requirements. The appropriateness of allocation of the animals to the existing feeding groups needs a continuous controlling. In the case of combined feeding, allocation of concentrate should be controlled and the successive adaptation process improved. Last but not least, animals which show clinical or subclinical signs of diseases and thus an overstressed capacity to adapt to the specific living conditions should be separated into a risk group given special attention and care. Nutrient and energy supply of endangered animals should be improved as well as their protection against biotic and abiotic stressors. Where appropriate, a temporary reduction in the outflow of glucose via the mammary gland should be considered to increase the availability of glucose in the body pool.

The overall production goal of farm management should be reoriented in striving for a prevalence of productions diseases that lies below the average of comparable dairy farms while simultaneously keeping the performance level on a level that does not compromise health and welfare of the farm animals. This goal cannot be achieved by general recommendations in relation to breeding and/or feeding. The history of Pds in dairy farming has proven that the predominant approach of animal science, based primarily on 'disposal knowledge' has failed to improve the long-lasting problems in relation to metabolic disorders and associated comorbidities. To solve problems which derive on different scales from very complex interactions between various factors and being to a high degree context-dependent requires also orientation and 'action knowledge'. Currently, many dairy farmers place their hopes in the development of further tools in precision dairy farming. This may certainly extend the options for acquiring more data but simply acquiring more data is not synonymous with gaining better 'action knowledge'. Before data can become farm-inherent knowledge for practical implementations, data need to be interpreted and transferred into valid information. Expectations, not least promoted by animal scientists, that the elaborate process can be automatized have failed so far. Not that there will be no further developments in the future and possible improvements but farm animals have already suffered too long. Farm management should not continue waiting for what might have little chance of effectiveness due to the underlying complexity of the processes. Whatever the future holds, there is first and foremost the necessity to solve problems in the here and now and to provide evidence of success in reducing production diseases, i.e., the implementation of measures in the farm-specific context together with external validation.

Author details

Albert Sundrum
Department of Animal Nutrition and Animal Health, University of Kassel, Witzenhausen, Germany

*Address all correspondence to: sundrum@uni-kassel.de

IntechOpen

© 2019 The Author(s). Licensee IntechOpen. This chapter is distributed under the terms of the Creative Commons Attribution License (http://creativecommons.org/licenses/by/3.0), which permits unrestricted use, distribution, and reproduction in any medium, provided the original work is properly cited.

References

[1] Knaus W. Dairy cows trapped between performance demands and adaptability. Journal of the Science of Food and Agriculture. 2009;**89**: 1107-1114. DOI: 10.1002/jsfa.3575

[2] LeBlanc S. Managing critical periods - transition dairy cows. In: 15th International Conference on Production Diseases in Farm Animals: book of abstracts; ICPD 2013. Uppsala, Sweden; 24-28 June 2013; 2013. pp. 62-65

[3] Krieger M, Sjöström K, Blanco-Penedo I, Madouasse A, Duval JE, Bareille N, et al. Prevalence of production disease related indicators in organic dairy herds in four European countries. Livestock Science. 2017;**198**:104-108. DOI: 10.1016/j.livsci.2017.02.015

[4] van Soest FJS, Mourits MCM, Blanco-Penedo I, Duval J, Fall N, Krieger M, et al. Farm-specific failure costs of production disorders in European organic dairy herds. Preventive Veterinary Medicine. 2019;**168**:19-29. DOI: 10.1016/j.prevetmed.2019.03.029

[5] Joshi NP, Herdt TH. Production Diseases in Farm Animals. The Netherlands: Wageningen Academic Publishers; 2006. DOI: 10.3920/978-90-8686-571-0

[6] Ametaj BN. Metabolic disorders of dairy cattle. Veterinary Science. 2010:56

[7] Mulligan FJ, Doherty ML. Production diseases of the transition cow. Veterinary Journal. 2008;**176**:3-9. DOI: 10.1016/j.tvjl.2007.12.018

[8] Oetzel GR. Undertaking nutritional diagnostic investigations. The Veterinary Clinics of North America: Food Animal Practice. 2014;**30**:765-788. DOI: 10.1016/j.cvfa.2014.08.002

[9] Herdt TH. Metabolic diseases of dairy cattle. The Veterinary Clinics of North America: Food Animal Practice. 2013;**29**:xi-xii. DOI: 10.1016/j.cvfa.2013.05.001

[10] Sundrum A. Metabolic disorders in the transition period indicate that the dairy cows' ability to adapt is overstressed. Animals. 2015;**5**:978-1020. DOI: 10.3390/ani5040395

[11] Drackley JK, Cardoso FC. Prepartum and postpartum nutritional management to optimize fertility in high-yielding dairy cows in confined TMR systems. Animal. 2014;**8**(s1):5-14. DOI: 10.1017/S175131114000731

[12] Bell AW. Regulation of organic nutrient metabolism during transition from late pregnancy to early lactation. Journal of Animal Science. 1995;**73**:2804-2819. DOI: 10.2527/1995.7392804x

[13] Sordillo LM, Raphael W. Significance of metabolic stress, lipid mobilization, and inflammation on transition cow disorders. The Veterinary Clinics of North America. Food Animal Practice. 2013;**29**:267-278. DOI: 10.1016/j.cvfa.2013.03.002

[14] van Saun RJ, Sniffen CJ. Transition cow nutrition and feeding management for disease prevention. The Veterinary Clinics of North America. Food Animal Practice. 2014;**30**:689-719. DOI: 10.1016/j.cvfa.2014.07.009

[15] Grant RJ, Albright JL. Feeding behavior and management factors during the transition period in dairy cattle. Journal of Animal Science. 1995;**73**:2791-2803. DOI: 10.2527/1995.7392791x

[16] Meyer U, Horstmann K, Kaske M, Flachowsky G. Inter-and intraindividual variation of feed intake and metabolic parameters of dairy cows related to energy supply. Publication European

Association for Animal Production. 2007;**124**:571

[17] Allen MS, Piantoni P. Metabolic control of feed intake: Implications for metabolic disease of fresh cows. The Veterinary Clinics of North America: Food Animal Practice. 2013;**29**: 279-297. DOI: 10.1016/j.cvfa. 2013.04.001

[18] Tylutki TP, Fox DG, Durbal VM, Tedeschi LO, Russell JB, van Amburgh ME, et al. Cornell net carbohydrate and protein system: A model for precision feeding of dairy cattle. Animal Feed Science and Technology. 2008;**143**:174-202. DOI: 10.1016/j.anifeedsci.2007.05.010

[19] Jensen LM, Nielsen NI, Nadeau E, Markussen B, Nørgaard P. Evaluation of five models predicting feed intake by dairy cows fed total mixed rations. Livestock Science. 2015;**176**:91-103. DOI: 10.1016/j.livsci.2015.03.026

[20] Allen MS, Bradford BJ, Oba M. Board invited review: The hepatic oxidation theory of the control of feed intake and its application to ruminants. Journal of Animal Science. 2009;**87**:3317-3334. DOI: 10.2527/ jas.2009-1779

[21] Corning PA, Kline SJ. Thermodynamics, information and life revisited, part I: 'To be or entropy'. Systems Research and Behavioral Science. 1998;**15**:273-295. DOI: 10.1002/(SICI)1099-1743(199807/08)15:4<273

[22] Weiner J. Physiological limits to sustainable energy budgets in birds and mammals: Ecological implications. Trends in Ecology and Evolution. 1992;**7**:384-388. DOI: 10.1016/0169-5347(92)90009-Z

[23] Glazier DS. Resource allocation patterns. In: Rauw WM, editor. Resource Allocation Theory Applied to Farm Animal Production. Wallingford: CABI; 2008. pp. 22-43. DOI: 10.1079/9781845933944.0022

[24] Rauw WM, editor. Resource Allocation Theory Applied to Farm Animal Production. Wallingford: CABI; 2008. DOI: 10.1079/9781845933944.0000

[25] Whittemore CT. Allocation of resources to growth. In: Rauw WM, editor. Resource Allocation Theory Applied to Farm Animal Production. Wallingford: CABI; 2008. pp. 130-146. DOI: 10.1079/9781845933944.0130

[26] Perotto D, Cue RI, Lee AJ. Comparison of nonlinear functions for describing the growth curve of three genotypes of dairy cattle. Canadian Journal of Animal Science. 1992;**72**: 773-782. DOI: 10.4141/cjas92-089

[27] Thorup VM, Højsgaard S, Weisbjerg MR, Friggens NC. Energy balance of individual cows can be estimated in real-time on farm using frequent liveweight measures even in the absence of body condition score. Animal: An International Journal of Animal Bioscience. 2013;**7**:1631-1639. DOI: 10.1017/S1751731113001237

[28] McEwen BS. Stress, adaptation, and disease. Allostasis and allostatic load. Annals of the New York Academy of Sciences. 1998;**840**:33-44

[29] Colditz IG. Allocation of resources to immune responses. In: Rauw WM, editor. Resource Allocation Theory Applied to Farm Animal Production. Wallingford: CABI; 2008. pp. 192-209. DOI: 10.1079/9781845933944.0192

[30] Aitken SL, Corl CM, Sordillo LM. Immunopathology of mastitis: Insights into disease recognition and resolution. Journal of Mammary Gland Biology and Neoplasia. 2011;**16**:291-304. DOI: 10.1007/s10911-011-9230-4

[31] Loiselle MC, Ster C, Talbot BG, Zhao X, Wagner GF, Boisclair YR, et al. Impact of postpartum milking frequency on the immune system and the blood metabolite concentration of dairy cows. Journal of Dairy Science. 2009;**92**:1900-1912. DOI: 10.3168/jds.2008-1399

[32] Esposito G, Irons PC, Webb EC, Chapwanya A. Interactions between negative energy balance, metabolic diseases, uterine health and immune response in transition dairy cows. Animal Reproduction Science. 2014;**144**:60-71. DOI: 10.1016/j.anireprosci.2013.11.007

[33] Wolowczuk I, Verwaerde C, Viltart O, Delanoye A, Delacre M, Pot B, et al. Feeding our immune system: Impact on metabolism. Clinical and Developmental Immunology. 2008;**2008**:639803. DOI: 10.1155/2008/639803

[34] Ingvartsen KL, Moyes KM. Factors contributing to immunosuppression in the dairy cow during the periparturient period. The Japanese Journal of Veterinary Research. 2015;**63**(Suppl 1): S15-S24

[35] Abuajamieh M, Kvidera SK, Fernandez MVS, Nayeri A, Upah NC, Nolan EA, et al. Inflammatory biomarkers are associated with ketosis in periparturient Holstein cows. Research in Veterinary Science. 2016;**109**:81-85. DOI: 10.1016/j.rvsc.2016.09.015

[36] Horst EA, Kvidera SK, Dickson MJ, McCarthy CS, Mayorga EJ, Al-Qaisi M, et al. Effects of continuous and increasing lipopolysaccharide infusion on basal and stimulated metabolism in lactating Holstein cows. Journal of Dairy Science. 2019;**102**: 3584-3597. DOI: 10.3168/jds.2018-15627

[37] Kvidera SK, Horst EA, Abuajamieh M, Mayorga EJ, Fernandez MVS, Baumgard LH. Glucose requirements of an activated immune system in lactating Holstein cows. Journal of Dairy Science. 2017;**100**:2360-2374. DOI: 10.3168/jds.2016-12001

[38] Larsen M, Kristensen NB. Effect of abomasal glucose infusion on splanchnic and whole-body glucose metabolism in periparturient dairy cows. Journal of Dairy Science. 2009;**92**:1071-1083. DOI: 10.3168/jds.2008-1453

[39] Baumgard LH, Collier RJ, Bauman DE. A 100-year review: Regulation of nutrient partitioning to support lactation. Journal of Dairy Science. 2017;**100**:10353-10366. DOI: 10.3168/jds.2017-13242

[40] Friggens NC, Brun-Lafleur L, Faverdin P, Sauvant D, Martin O. Advances in predicting nutrient partitioning in the dairy cow: Recognizing the central role of genotype and its expression through time. Animal: An International Journal of Animal Bioscience. 2013;7(Suppl 1): 89-101. DOI: 10.1017/S1751731111001820

[41] Patton J, Kenny DA, Mee JF, O'Mara FP, Wathes DC, Cook M, et al. Effect of milking frequency and diet on milk production, energy balance, and reproduction in dairy cows. Journal of Dairy Science. 2006;**89**:1478-1487. DOI: 10.3168/jds.S0022-0302(06)72215-9

[42] Melendez P, Marin MP, Robles J, Rios C, Duchens M, Archbald L. Relationship between serum nonesterified fatty acids at calving and the incidence of periparturient diseases in Holstein dairy cows. Theriogenology. 2009;**72**:826-833. DOI: 10.1016/j.theriogenology.2009.06.001

[43] Ospina PA, Nydam DV, Stokol T, Overton TR. Evaluation of nonesterified

fatty acids and beta-hydroxybutyrate in transition dairy cattle in the northeastern United States: Critical thresholds for prediction of clinical diseases. Journal of Dairy Science. 2010;**93**:546-554. DOI: 10.3168/jds.2009-2277

[44] Sordillo LM, Mavangira V. The nexus between nutrient metabolism, oxidative stress and inflammation in transition cows. Animal Production Science. 2014;**54**:1204. DOI: 10.1071/AN14503

[45] Gardner GE, McGilchrist P, Pethick DW. Ruminant glycogen metabolism. Animal Production Science. 2014;**54**:1575. DOI: 10.1071/AN14434

[46] Aschenbach JR, Kristensen NB, Donkin SS, Hammon HM, Penner GB. Gluconeogenesis in dairy cows: The secret of making sweet milk from sour dough. IUBMB Life. 2010;**62**:869-877. DOI: 10.1002/iub.400

[47] Larsen M, Kristensen NB. Precursors for liver gluconeogenesis in periparturient dairy cows. Animal. 2013;**7**(10):1640-1650. DOI: 10.1017/S1751731113001171

[48] Bauman DE, Currie WB. Partitioning of nutrients during pregnancy and lactation: A review of mechanisms involving homeostasis and homeorhesis. Journal of Dairy Science. 1980;**63**:1514-1529

[49] Sterling P, Eyer J. Allostasis: A new paradigm to explain arousal pathology. In: Fisher S, Reason J, editors. Handbook of Life Stress, Cognition and Health and health. Oxford, England: John Wiley & Sons; 1988. pp. 629-649

[50] Erdman RA, Varner M. Fixed yield responses to increased milking frequency. Journal of Dairy Science. 1995;**78**:1199-1203. DOI: 10.3168/jds.S0022-0302(95)76738-8

[51] Stefanon B, Colitti M, Gabai G, Knight CH, Wilde CJ. Mammary apoptosis and lactation persistency in dairy animals. The Journal of Dairy Research. 2002;**69**:37-52

[52] Bauman DE, Mather IH, Wall RJ, Lock AL. Major advances associated with the biosynthesis of milk. Journal of Dairy Science. 2006;**89**:1235-1243. DOI: 10.3168/jds.S0022-0302(06)72192-0

[53] Eastridge ML. Major advances in applied dairy cattle nutrition. Journal of Dairy Science. 2006;**89**:1311-1323. DOI: 10.3168/jds.S0022-0302(06)72199-3

[54] Lucy MC. Mechanisms linking nutrition and reproduction in postpartum cows. Reproduction Supplement. 2003;**61**:415-427

[55] Sterling P. Allostasis: A model of predictive regulation. Physiology and Behavior. 2012;**106**:5-15. DOI: 10.1016/j.physbeh.2011.06.004

[56] Friggens NC, Newbold JR. Towards a biological basis for predicting nutrient partitioning: The dairy cow as an example. Animal: An International Journal of Animal Bioscience. 2007;**1**:87-97. DOI: 10.1017/S1751731107657772

[57] de Vries MJ, van der Beek S, Kaal-Lansbergen LMTE, Ouweltjes W, Wilmink JBM. Modeling of energy balance in early lactation and the effect of energy deficits in early lactation on first detected Estrus postpartum in dairy cows. Journal of Dairy Science. 1999;**82**:1927-1934. DOI: 10.3168/jds.S0022-0302(99)75428-7

[58] Beerda B, Ouweltjes W, Šebek LBJ, Windig JJ, Veerkamp RF. Effects of genotype by environment interactions on milk yield, energy balance, and protein balance. Journal of Dairy Science. 2007;**90**:219-228. DOI: 10.3168/jds.S0022-0302(07)72623-1

[59] Kessel S, Stroehl M, Meyer HHD, Hiss S, Sauerwein H, Schwarz FJ, et al. Individual variability in physiological adaptation to metabolic stress during early lactation in dairy cows kept under equal conditions. Journal of Animal Science. 2008;**86**:2903-2912. DOI: 10.2527/jas.2008-1016

[60] van der Drift SGA, Jorritsma R, Schonewille JT, Knijn HM, Stegeman JA. Routine detection of hyperketonemia in dairy cows using Fourier transform infrared spectroscopy analysis of β-hydroxybutyrate and acetone in milk in combination with test-day information. Journal of Dairy Science. 2012;**95**:4886-4898. DOI: 10.3168/jds.2011-4417

[61] Hattan AJ. Energy Utilisation in High Yielding Dairy Cows. PhD Thesis. UK: University of Reading; 2003

[62] van Knegsel ATM, van den Brand H, Dijkstra J, van Straalen WM, Jorritsma R, Tamminga S, et al. Effect of glucogenic vs. lipogenic diets on energy balance, blood metabolites, and reproduction in primiparous and multiparous dairy cows in early lactation. Journal of Dairy Science. 2007;**90**:3397-3409. DOI: 10.3168/jds.2006-837

[63] Razeto-Barry P. Autopoiesis 40 years later. A review and a reformulation. Origins of life and evolution of the biosphere: The Journal of the International Society for the Study of the Origin of Life. 2012;**42**: 543-567. DOI: 10.1007/s11084-012-9297-y

[64] Savietto D, Friggens NC, Pascual JJ. Reproductive robustness differs between generalist and specialist maternal rabbit lines: The role of acquisition and allocation of resources. Genetics, Selection, Evolution. 2015;**47**: 2. DOI: 10.1186/s12711-014-0073-5

[65] Ingvartsen KL, Dewhurst RJ, Friggens NC. On the relationship between lactational performance and health: Is it yield or metabolic imbalance that cause production diseases in dairy cattle? A position paper. Livestock Production Science. 2003;**83**:277-308. DOI: 10.1016/S0301-6226(03)00110-6

[66] Doehring C, Sundrum A. The informative value of an overview on antibiotic consumption, treatment efficacy and cost of clinical mastitis at farm level. Preventive Veterinary Medicine. 2019;**165**:63-70. DOI: 10.1016/j.prevetmed.2019.02.004

[67] Krieger M, Hoischen-Taubner S, Emanuelson U, Blanco-Penedo I, de Joybert M, Duval JE, et al. Capturing systemic interrelationships by an impact analysis to help reduce production diseases in dairy farms. Agricultural Systems. 2017;**153**:43-52. DOI: 10.1016/j.agsy.2017.01.022

[68] Krieger M, Schwabenbauer E-M, Hoischen-Taubner S, Emanuelson U, Sundrum A. Graph-based impact analysis as a framework for incorporating practitioner knowledge in dairy herd health management. Animal. 2018;**12**:624-633. DOI: 10.1017/S1751731117002105

[69] McNamara S, Murphy JJ, O'Mara FP, Rath M, Mee JF. Effect of milking frequency in early lactation on energy metabolism, milk production and reproductive performance of dairy cows. Livestock Science. 2008;**117**:70-78. DOI: 10.1016/j.livsci.2007.11.013

[70] Wathes DC, Cheng Z, Chowdhury W, Fenwick MA, Fitzpatrick R, Morris DG, et al. Negative energy balance alters global gene expression and immune responses in the uterus of postpartum dairy cows. Physiological Genomics. 2009;**39**:1-13. DOI: 10.1152/physiolgenomics.00064.2009

[71] Ster C, Loiselle M-C, Lacasse P. Effect of postcalving serum nonesterified fatty acids concentration on the functionality

of bovine immune cells. Journal of Dairy Science. 2012;**95**:708-717. DOI: 10.3168/jds.2011-4695

[72] Soberon F, Ryan CM, Nydam DV, Galton DM, Overton TR. The effects of increased milking frequency during early lactation on milk yield and milk composition on commercial dairy farms. Journal of Dairy Science. 2011;**94**:4398-4405. DOI: 10.3168/jds.2010-3640

[73] Phyn CV, Kay JK, Rius AG, Davis SR, Stelwagen K, Hillerton JE, et al. Review: Impact of Short-Term Alterations to Milking Frequency in Early Lactation. New Zealand: Caxton Press; 2010

[74] Stelwagen K, Phyn CVC, Davis SR, Guinard-Flament J, Pomies D, Roche JR, et al. Invited review: Reduced milking frequency: Milk production and management implications. Journal of Dairy Science. 2013;**96**:3401-3413. DOI: 10.3168/jds.2012-6074

[75] Sundrum A. Healthy food from healthy cows. In: Konvalina P, editor. Organic Farming and Food Production. Rijeka, Croatia: IntechOpen; 2012. DOI: 10.5772/53150

[76] Sundrum A. Assessment of animal protection services in livestock farming. Berichte über Landwirtschaft - Zeitschrift für Agrarpolitik und Landwirtschaft. 2018;**96**:1-33. DOI: 10.12767/BUEL.V96I1.189

[77] Sundrum A. Real-farming emissions of reactive nitrogen—Necessities and challenges. Journal of Environmental Management. 2019;**240**:9-18. DOI: org/10.1016/j.jenvman.2019.03.080

[78] Rutten CJ, Velthuis AGJ, Steeneveld W, Hogeveen H. Invited review: Sensors to support health management on dairy farms. Journal of Dairy Science. 2013;**96**:1928-1952. DOI: 10.3168/jds.2012-6107

Chapter 5

Nutrients Cycle within Swine Production: Generation, Characteristics, Treatment and Revaluation

Cristina Alejandra Villamar and Cristóbal Sardá

Abstract

The swine production generates slurries nutrients rich, which could be revaluated in cereal crops used for its food and energy generation (biogas) for use on the farm. However, the revaluation requires to know their physical–chemical and biological characteristics, which allow giving an adequate transformation (treatment). On the one hand, swine production and consumption market reveal the superiority of emergent countries on meat/cereal (feed) production and swine meat consumption (concentrated population). The food composition and growth phase will influence the swine slurries composition, which is rich in organic matter, macronutrients (N, P) and micronutrients (Cu and Zn). These characteristics will generate odors (organic matter, macronutrients) and ecotoxicology effects (macro/micronutrients) if they are not treated. Moreover, the swine slurries treatment allows revaluated them in agriculture and obtaining energy. Anaerobic technologies (anaerobic lagoon, mixed complete reactors, UASB, among others) are the most used/cost-effective to organic matter removal from swine slurries, obtaining from 0.28 to 0.83 m^3 biogas/kg organic matter. Meanwhile, passive technologies (constructed wetlands) are the most used technologies to nutrients and metals removal. Treated swine slurries from constructed wetlands have agronomic properties. Therefore, the nutrients cycle within swine production would favor concepts of revaluation in origin.

Keywords: biogas, soil fertilizer, nutrients, revaluation, swine slurries

1. Introduction

The economic and demographic growths are the factors that are activating the current world meat production [1]. Thus, developing countries where 75% of the population is located also concentrate 52% (41.5 million ton/year) of swine production. Moreover, in the last 40 years the growth of emergent economies are concentrating both meat and cereal production in one place. China is an emergent economy in where is 46% of the world's swine production, and 30% of cereal production.

On the other hand, demand is the main trigger factor of the current "livestock revolution" [2]. Thus, the meat consumption acquires connotation different depending on the consumer type. In developed countries, higher purchasing power

IntechOpen

and demanding lifestyles have promoted a higher swine meat demand (25 kg/per capita year). Meanwhile, developing countries consume only 8 kg/per capita year, condition related mainly with subsistence habits. However, developing countries (75% worldwide population) concentrates 64% (43.3 million ton/year) of the swine meat total consumption; while that, developed countries concentrate 36% (34.4 million ton/year) remaining. However, emergency economies (Brazil and China) are increasing about 21% the swine meat consumption per capita of their population, concentrating swine meat consumption in these countries [3].

2. Swine slurry: generation and characteristics

2.1 Origin of swine slurry

The swine slurry generation is related to the growth phase and its water/food requirements. Thus, lactation and reproduction phases have a great water supply (12.2–41.1 L water/animal d), mainly related to hydration improvement or fertility [4, 5]. Meanwhile, fattening and weaning have a greater food intake (1.9 kg food/animal d), exclusively given by increased weight or age [5, 6].

The swine diet is based on proteins, carbohydrates and starch concentrates, which exceed 50% of food total composition [7]. Indeed, fattening tends to prioritize the protein intake (>30%) [6]; while, reproduction and weaning consume mainly fiber (10%) to avoid overweight [8]. In both cases, the low food digestibility could generate nitrogen excretion (<30%) and phosphorus (<10%) [9]. This factor can be observed in the swine conversion index, which can vary from 2.0 to 4.3 and 0.9 to 2.3 kg feed/kg weight gained for fattening and previous phases, respectively [6, 10–12]. Indeed, digestibility can vary depending on the food and animal type, being slightly lowers in the final growth phases (fattening ~ 25%) than initial phases (weaning >50%).

Finally, the water/food ratios allow indirectly evaluating the swine slurry quantity and its composition. Thereby, initial phases (weaning) generate the water/feed ratios from 4.4 to 5.1 L water/kg food [8]; while, final phases (fattening) reaches values from 1.6 to 2.5 L water/kg food [11]. This indicator is indirectly related with nitrogen concentrations from urine/feces values. Indeed, greater urine/feces values are obtained by fattening (6.6–9.1) than weaning (5.5–6.3) [13]. At this point, it's important to note that within the swine excreta around 75% of the nitrogen is in urine, so a higher water intake will increase the nitrogen generation in the slurries [14] (**Table 1**).

2.2 Generation and physicochemical characteristics

The physicochemical composition from swine slurries will be influenced by the low digestibility of nitrogen (<33%), phosphorus (<32%) and micronutrients (<3%) [9, 17], as well as the animal growth phase [18]. Initial phases (maternity and lactation) produce more slurries (10.0–41.1 L/animal d) than weaning-fattening (3.5–9.10 L/animal d) [4–6]. This could be related with dilution by more water intake. **Table 2** summarizes the physicochemical characteristics from swine slurries according to the growth phase. The main differences are related to a higher content of organic matter (1.8–2.4 times), nitrogen (1.3–2.8 times) and phosphorus (2.1–4.6 times) from fattening with respect to maternity and weaning. This condition is related to a higher food intake (1.7 times), which is made up of carbohydrates and proteins [14]. Meanwhile, micronutrients, such as zinc are excreted (2.7–13.0 times) in the initial phases (weaning). In a different proportion, copper is excreted by swine slurries, increasing during fattening (1.5–7.2 times). Differences in the

Feed type	Conversion index kg feed/kg weight gained	Water/food ratio L water/kg feed	Reference
Pellet	2.13–3.33	—	[15]
With/out phytases	2.00–4.32	—	[6]
Wheat and soybeans (12–20%)	2.22–2.35	—	
Intensive/organic	2.70–3.20	—	[10]
Ad libitum	2.95–3.05	1.99–2.60	[11]
Wean/mat/fat food	0.98–2.14	2.00–20.00	[12, 16]

Table 1.
Productive indicators within swine production.

		Growth phase			
Parameter	**Unit**	**Maternity**	**Weaning**	**Fattening**	**Reference**
Slurry	L/animal d	10.0–16.0	23.5–41.1	3.5–9.1	[4–6]
pH		7.5	6.9	7.2–8.4	[6, 18, 20–22]
EC	mS/cm	12.8–15.5	14.2	15.3–25.3	[18, 21, 22, 24]
TS	%	1.7	2.7	3.2	[18, 19, 24]
BOD_5	g/L	9.0	25.0	16.6–21.6	[18, 22, 24, 25]
COD	g/L	24.0	65.2	45.3–57.7	[6, 18, 22, 24, 25]
TN	g/L	1.8–2.3	2.3	2.4–4.6	[6, 18–22, 24, 25]
$N\text{-}NH_4^+$	g/L	1.4–1.8	1.5	2.0–3.1	[6, 18–22, 24, 25]
TP	g/L	1.4	0.6	0.8–2.8	[18–22]
K	g/L	2.2	1.8	1.9–3.8	[18–22]
Cl^-	g/L	3.6	2.3	5.1	[18, 24]
Cu	mg/L	11.0	55.0	15.9–80.0	[18, 22, 24, 26]
Zn	mg/L	75.0	533.0	40.7–191.0	[18, 22, 24, 26]

Table 2.
Physicochemical characteristics from swine slurries.

presence of copper and zinc excreted is related to the use of both metals as growth promoters and specifically copper that is used additionally for therapeutic purposes (fighting diarrhea) in the most vulnerable swine population (weaning) [17]. Chloride, ammonium and potassium salts may also be present within swine slurries. The feeding has influenced a greater salts excretion (1.4–2.3 times) during weaning-fattening. These characteristics are related to the fact that the salts in the swine feeding, favor the liquids retention increasing weight during the growth [18].

On the other hand, raw swine slurries show agronomic properties (N:P: K between 1.1:0.6:1.0 and 1.3:0.4:1.0) [18–22]. Indeed, the nutritional requirements of cereal crops (e.g. corn, wheat) can reach values of N:P: K between 1.2:0.2:1.0 and 1.6:0.3:1.0 [23]. However, the presence of pathogenic organisms and micro-contaminating (metals, antibiotics) may limit their use prior to treatment.

2.3 Ecotoxicological characteristics

Swine slurries have ecotoxicological characteristics depending on the bioindicator used. **Tables 3** and **4** summarize the ecotoxicological characteristics (swine slurries and composition) on terrestrial and aquatic bio-indicators. The ecotoxicological

Organism	Test	Compound	Concentration (mg/L o %*)	Reference
Swine slurries				
Daphnia magna	48–24 h-LC_{50}	Raw swine slurry	1.8–5.0	[27, 28]
Nutrients				
Moina macrocopa	24 h-LC_{50}	NH_4^+	231–492	[31]
Penaeus semisulcatus	96 h-LC_{50}	NH_4^+	11.4–55.9	[36]
Ceriodaphnia dubia	48 h-LC_{50}	NH_4^+	20	[32]
Daphnia carinata	24–48 h-LC_{50}	NH_4^+	2.2–2.8	[33]
Monia australiensis	24–48 h-LC_{50}	NH_4^+	7.5–8.5	[33]
Daphnia magna	48 h-LC_{50}	NH_3	2.9–6.9	[34]
Metals				
Daphnia magna	48 h-LC_{50}	Zn	0.05	[41]
Daphnia magna	48 h-LC_{50}	Cu	0.56	[41]
Daphnia magna	LOEC 14d	Cu, Zn	0.12	[44]
Danio rerio	96 h-LC_{50}	Cu	0.12–0.13	[41]
Salts				
Oncorhynchus mykiss	6 h-LC_{40}	NaCl	20,000	[39]
Daphnia magna	24-48 h-LC_{50}	NaCl	1020–3240	[31]
Daphnia magna	48 h-LC_{50}	Cl_2	0.1–0.2	[34]
Daphnia pulex	48 h-LC_{50}	Cl^-	2042	[45]

% = percentage concentration.*

Table 3.
Ecotoxicological characteristics on aquatic organisms.

Organism	Test	Compound	Concentration (mg/L o %*)	Reference
Swine purines				
Lepidium sativum L.	Growth 24, 48, 72 h	raw swine slurry	3–10*	[29]
Eisenia foetida	Growth 28d	Untreated purine	25*	[30]
Nutrients				
Lactuca sativa	Growth 7d	NH_3	0.002	[38]
Hordeum vulgare	Growth 8d	NH_4^+	0.18	[37]
Metals				
Eisenia foetida	48 h-LC_{50}	Zn	268–439	[43]
Eisenia foetida	48 h-LC_{50}	Cu	153–249	[43]
Salts				
Different kind of plants	Growth	NaCl	0–1280	[40]

% = percentage concentration.*

Table 4.
Ecotoxicological characteristics terrestrial organisms.

studies have been carried out mainly in *Daphnia magna*, establishing higher acute toxicity at low concentrations (48 h-LC_{50}, 1.8–3.3%) [27, 28]. Meanwhile, chronic ecotoxicity has been observed on *Lepidium sativum* L. (growth inhibited at concentrations from 3 to 10%, v/v) [29]. However, *Eisenia foetida* growth (6000–20,000 worms/m^2) has also been reported at concentrations less than 25% [30]. Nutrients, salts, metals, antibiotics, among others separately can generate effect at different levels. On the one hand, NH_4^+ within swine slurries reaches values above 1 g/L [18]. Thus, ammonium on cladoceran aquatic organisms (*Ceriodaphnia dubia, Moina macrocopa, Daphnia carinata, Monia australiensis*) generate mortality at values above 7.5 mg NH_4^+/L [31–33]. Indeed, ammonium causes enzymatic inhibition and cell disruption [34]. Moreover, NH_4^+ is instable transforming to NH_3 under alkaline pH (>8) and/or temperature increasing [35]. Indeed, *Daphnia magna* shows acute toxicity at ammonia concentrations above 2.9 mg/L [34]. Meanwhile, organisms of higher trophic levels (*Penaeus semisulcatus*) reports acute toxicity at concentrations from 11 to 55 mgNH_4^+/L [36]. In the soil, toxic effects of NH_4^+/NH_3 have been observed mainly in vegetable species, generating necrosis, reduction/stimulation of growth and sensitivity to frost [38]. Indeed, *Lactuca sativa* and *Hordeum vulgare* report chronic effects (growth decreasing) at concentrations from 0.002 mg NH_3/L [37, 38].

Chlorides can exceed 1 g/L within swine slurries, being reported less acute toxicity of NaCl (1020–3240 mg/L) on *Daphnia magna* than Cl_2 (0.1–0.2 mg/L) [18, 31, 34]. However, toxic effect is decreased at high trophic levels (the chlorinated compounds toxicity decreases on higher trophic levels (*Oncorhynchus mykiss*)), reaching concentrations above 20 g/L [39]. In terrestrial environments, chlorides can cause chronic effects on the vegetal germination [40]. Swine slurries are compound also by micronutrients (copper and zinc), which not exceeds 0.5 g/L [26]. However, acute toxicity on aquatic organisms (*Danio rerio*) reaches concentrations about 0.1 mg/L and on terrestrial (*Eisenia foetida*) it is not exceeding at values of 249 mg/L [41–43].

2.4 Olfatometric characteristics

The odoriferous characteristics within swine slurries are evaluated using analytical methods (compounds) and sensory methods (odor) [46]. Olfactometric characteristics measured as odor concentration (OC/m^3) are influenced by the type/time of slurries storage. [47]. Indeed, fattening phase (2.1–5.6 times) has more odor than maternity/weaning. On the other hand, fresh swine slurries (1.5–5.4 times) generate more odor than stored slurries (2 months). These results have also shown the odor relationship with volatile compounds presence, which from by diet or slurries management. Thereby, sulfur, ammonia, phenolic and volatile fatty acid compounds from swine slurries influence the odor generation, which show different correlation degree (R^2 = 0.66–0.89) with the generated odor [46]. Particularly, odor precursor compounds from the swine production will be given by the protein's presence from food, generating sulfurous, indole, phenolic and long-chain fatty acids [47–49]. Meanwhile, carbohydrates generate short-chain or volatile fatty acids (less than 6 carbons) [40]. Thus, biological and chemical conditions determine the odor compounds formation. It has been possible to establish the presence of some autochthonous microorganisms (*Streptococcus, Peptostreptococcus, Eubacterium, Lactobacilli, Escherichia, Clostridium, Propionibacterium, Bacteroides, Megasphaera*, among others). Incomplete anaerobic digestion from these microorganisms produce and/or reduce macromolecules to odor compounds, so intermediates (e.g. volatile fatty acids) and finals (e.g. sulfides) [50, 51]. However, not only microorganisms can condition the odorant compounds formation, but also environmental conditions (pH and temperature), which active biological and chemical

Type of waste	Olfactometric technique	Odor threshold (D-T)*, Odor concentration (OC/m³)^ or Odor index (OI°)	Reference
Bovine manure	Field (Nasal Ranger, box)	8.6–157.7*	[55]
	Dynamic	124.2–6561^	
Sheep, bovine and pig manure	Field (Nasal Ranger)	<2–60*	[56]
Swine purines	Dynamic	120–792^	[46, 47]
Swine manure	Triangular odor bags (dynamic mod.)	26.2–58.7°	[57]
Porcine, avian and bovine manure	Dynamic	<1–1000^	[58]

**=D-T units, ^ = OC/m³ units,° = OI units.*

Table 5.
Olfactometric characteristics from swine slurries.

(e.g. ammonium/ammonia) processes [52, 53]. Temperature affects the microbial growth rate; while, pH influences the buffer capacity, favoring volatile fatty acids generation [54] (**Table 5**).

3. Reduction and Re-valorization of swine slurries

3.1 Slurries reduction

Swine has low digestibility (<30%) of nutrients and micronutrients, being necessary mechanisms of digestibility improvement, which could improve the physicochemical characteristics from slurries (feces + urine) [9]. Indeed, about 78% of N and P from swine food (proteins) is not assimilated, excreting concentrated urine and feces [14]. Therefore, farm managements are focused on the improvement of diet type and food quantity during each phase growth. On the one hand, raw protein is substituted by fiber, reducing until 8% (10 g RP/kg food) of nitrogen in the urine [7]. Other strategies are related to vary crude protein concentration (155, 145 and 135 g RP/kg) in the food, achieving the decrease of NH_4^+ (20.3–28.4%) in the excreta [13]. Studies have evaluated the replacement of crude protein by digestible or ileal amino acids (lysine, threonine, methionine, tryptophan, isoleucine and valine), finding that they can reduce the ammonium excretion in the urine from 40 to 50% [69]. Meanwhile, other techniques use feeding multi-phases, which improve the protein digestibility, reducing between 20 and 42% the nitrogen excretion [59, 60]. On the other hand, introducing phytases in the swine diet, it is possible to reduce 18% of phosphorus in the feces. Metals (Cu, Zn) used as growth promoters have been decreased (100–250 ppm Cu, 2000–3000 ppm Zn) by antibiotics (3–220 g/ton food) [61, 62]. Indeed, sulfonamides, tetracyclines and β-lactams increase the index conversion rate between 3 and 4%, improving the protein assimilation [62]. However, antibiotics also are excreted up to 10%, not being a good strategy because they are emerging contaminants [63]. The implementation of efficient water drinking reduces the slurry generation. The dozers incorporation and excreta handling techniques (e.g. hot beds) could reduce the floor washing, reusing waste organics (rice husk, straw) [64]. These strategies have reached reduce water requirements from 5 to 80% [4].

3.2 Treatment and re-evaluation of by-products

The swine slurry treatment is the most used tool management within intensive farms by environmental pressures (legislation), which regulates its discharge on water bodies or soil revaluation. The slurries management requires a balance between the environmental/social and economic requirements in the farms. Ideally, this management starts with the excreta fractionation (slurry = urine + feces/feces) by physical/chemical separation. Some techniques, such as: polymers, filter press, flotation, sedimentation, screw press, among others are used to remove sedimentable/suspended material, reducing mainly organic matter (62–84%) and phosphorus (70–89%) [65]. The solid fraction corresponding to non-mineralized organic matter is subjected to composting (aerobic/anaerobic), which stabilizes giving it agronomic properties (C/N < 20). The liquid fraction (slurries) with a C/N ratio about 10 is subjected to biological (aerobic/anaerobic) removal processes of organic matter, nutrients and other microcontaminants (metals, emergent) [66]. Several technologies are grouped within the aerobic biological processes (aerated lagoons, activated sludge, among others) and anaerobic (anaerobic lagoons, fixed bed reactors, SBR or Sequencing Batch Reactor, UASB or Upflow Anaerobic Sludge Blanket, among others) [67]. Thus, the removal of dissolved and colloidal organic matter is usually carried out by anaerobic lagoons (*Environmentally Superior Technologies*), which reduce 50% of organic matter. Meanwhile, anaerobic reactors (Manure-based biogas plants) remove more than 80% of organic matter. The by-products obtained from this stage are usually stabilized effluents (C/N < 10) used as soil stabilizer and biogas [65, 68]. In this last point, specific temperature conditions (psychrophilic, mesophilic and thermophilic) have allowed the biogas (60–70% CH_4, 40–30% CO_2) production between 0.03 (anaerobic lagoons) and 650 (anaerobic reactors) m^3/d. The main biogas uses are related with thermal energy (0.02–390 m^3 gas) and/or electrical (0.07–1560 kWh) within farms [65, 69]. Additionally, anaerobic treatment under optimum conditions (35°C) reduces odors (1.9 units depending on the hedonic tone) [52]. This anaerobically treated effluent can be subjected to biological treatment (nitrification/denitrification, SBR or constructed wetlands) [70–72] or physical–chemical (stripping, vacuum evaporation, precipitation) [71, 73, 74] to nutrients removal (nitrogen and phosphorus). The nitrogen removal efficiencies vary from 40% (constructed wetlands) to 97% (denitrification–denitrification) and 100% (stripping) [71, 73]. Meanwhile, phosphorus is removed between 44% (constructed wetlands) and 80% (chemical precipitation) [71, 75–77]. In very few cases, have been reported metals removal (Cu, Zn), mainly due to their low concentrations (<1 g/L) [26]. However, the metals removal has allowed to obtain removal efficiencies between 75% (precipitation) and 92% (constructed wetlands) [78, 79]. The by-products obtained in this stage can vary from crystallized ammonium salts for agronomic use [73] to treated effluents with a C/N ratio <5 usable in irrigation [71].

The irrigation (slurries) or soil stabilization (solid) are the most used re-valuation techniques. The treated swine slurries have nutritional value (N:P:K: 1:0.6:0.4–1:0.3:1) to be used in cereals irrigation for swine consumption (1.2:0.2:1 a 1.6:0.3:1) [23, 80]. Under optimal irrigation conditions (150–200 kg N/ha year) some soil characteristics with agronomic importance (organic matter content and moisture retention) could be improved [81, 82]. The slurries re-valorization in irrigation has decreased the chemical fertilizers use, being in some countries (New Zealand) valued economically (21 million USD/year) [82]. However, the livestock production intensification vs. land availability (Europe) has carried out to optimize the nutrients recovery. Thus, it is necessary to consider within slurries the balance of macro (N:P: K) and micronutrients (metals), as well as other contaminants (pathogens, emergent). Moreover, this balance must consider soil nutritional requirements and

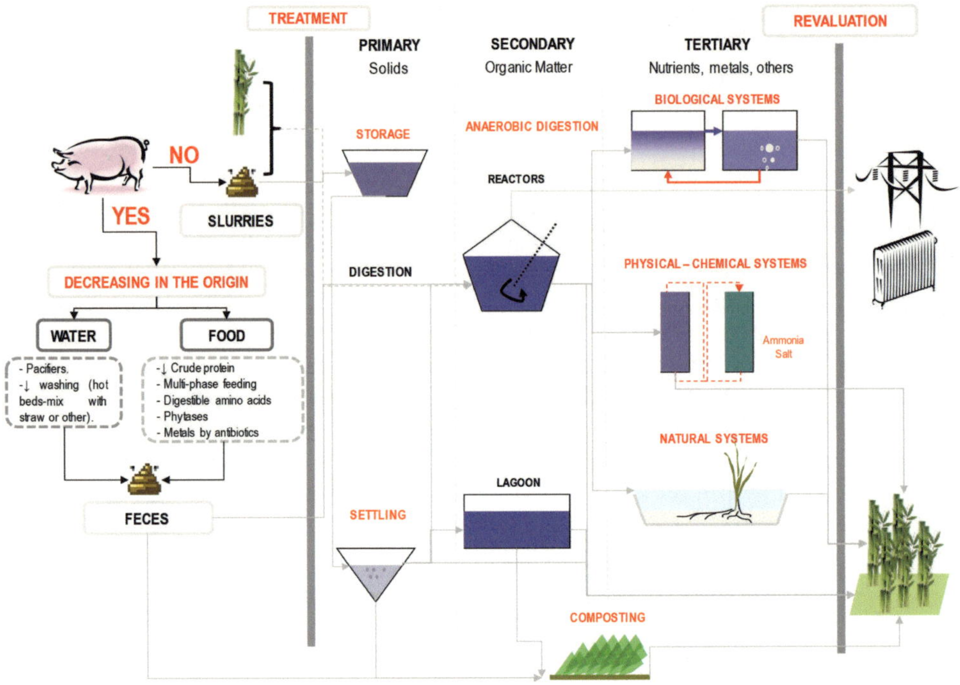

Figure 1.
Diagram of decision-making process within of the cycle of generation, treatment, and revaluation of swine slurry.

crop type, according to the international revaluation legislation [83, 84]. These tools would allow the cadasters establishment, which could facilitate the communication between nutritional requirements sites (agricultural soils) and nutrients-generating sites (animal farms) [85]. Preliminary Chilean studies have reported opportunities of livestock slurries revaluation in agriculture relating nutrients recovery sites with adequate agriculture sites [86].

The biogas production from anaerobic digestion of swine slurries is another revaluation alternative. In Europe, technologies based on biogas generation have been favored by state subsidies (10–30% investment cost) at farms level. These initiatives have allowed the building the more than 5000 anaerobic digestion projects [65, 87]. Swine slurries have potential capacity of methane generation between 0.25 and 0.30 m^3CH_4/kg VS [88]. Meanwhile, co-digestion with lignocellulosic materials (e.g. crop waste) could increase more than 16% produced methane [89]. Thus, centralized plants of anaerobic co-digestion are the current trend within European agricultural sector. Successful experiences have been reported in Germany and Denmark, where more than 32 plants have been built with a capacity between 16,000 and 200,000 tons/year of waste and with a production between 0.7 and 5.7 million m^3 biogas/year [90, 91]. Preliminary Chilean studies have allowed to establish the potential generation of bioenergy (biogas) from anaerobic co-digestion (livestock/crop wastes) at regional level [92]. **Figure 1** describes excretes management in the swine production.

4. Swine slurries treatment technologies

4.1 Organic matter removal

Anaerobic treatment technologies are widely used in this field, because they allow to remove organic matter, pathogens, odors, generating by-products (biogas, bio-fertilizer) [93]. There several anaerobic technologies, such as: biodigesters, anaerobic lagoon, complete mixing anaerobic reactors, anaerobic filters, UASB or Upflow Anaerobic Sludge Blanket, among other [94]. Worldwide, anaerobic technology based on conventional anaerobic mesophilic reactors (4935 plants) and thermophilic reactors (321 plants) treat around 49 million ton/year slurries, which have been built in Europe, mainly in Germany (more than 70%) [87]. State subsides, innovation technologic and centralization of biogas plants using agricultural wastes (livestock, crop) have favored their implementation [90, 91]. Meanwhile, more than 7000 technologies based on anaerobic lagoon have been built, but with only 60% operationally actives in the United States [63]. The conventional anaerobic reactors (complete mixture, piston flow) have been limited by its costs, remaining in disuse or only in building project around 77% [94]. However, currently these technologies are being replaced by more efficient technologies in terms of removal of nutrients, odors and pathogens [68, 71]. Technologies such as tubular biodigesters are the most used in countries where the livestock production in non-intensive.

The *anaerobic lagoons* are extensive and conventional typology, where organic matter is biodegraded without hydraulic or thermal control (environmental conditions) [95]. In the livestock sector, these systems offer some advantages related to their storage capacity and operation easy [96]. However, they can also cause odors, requiring spaces far spaces from neighboring population [94, 96]. Anaerobic lagoons obtain organic matter removal efficiencies between 26 and 79%, working under environmental conditions (5–32°C) and with residence times from 90 to 232 days [97, 98]. The lagoon design must consider thermal vertical stratification, suggesting depths between 0.8 and 4.0 m to maintain facultative conditions upper and anaerobic zone bottom [69, 99]. Its longer residence time also favors the macro (organic matter, nutrients) and micro (metals) nutrients precipitation. Thus, has been observed that anaerobic lagoon accumulates more than 50% COD, TN, P and Cu in the bottom, decreasing pH from 6.5–7.2 throughout its depth (>1.5 m) [99]. Another advantage is related with their disinfection capacity, due to its prolonged time exposure to solar radiation, causing cell lysis of pathogenic microorganisms [100]. However, this time exposure generates greenhouse gas emission (0.02–0.5 m^3 biogas/m^2 d) [69] and odors (168–262 OC/m^3, 101 μg NH_3/m^2 s, 5.7 μg H_2S/m^2 s) [101]. Operational improvements mainly related to the emission of gases from anaerobic lagoons have been made covering them.

The conventional full mix reactors or CSRT (Continuous Stirred Reactor Tank) are controlled systems, where the hydraulic retention time (residence time) is equal to the cell retention time [102]. The complete mixture is achieved through the recirculation from 25 to 40% of biogas generated [103]. It can operate under psychrophilic, mesophilic, and thermophilic conditions, which are carried out in two stages (acidogenic and methanogenic reactors) [93, 103]. This technology has been widely used mainly in Europe, obtaining organic matter removal efficiencies from 25 to 74%, but with organic loading between 5 and 40 times greater than anaerobic lagoon [87, 93]. CSRT reactor have been improved with the recirculation [93].

Other conventional technologies as anaerobic filters or AF generate biofilm around of the material support surface, while the flow goes up throughout the filter [102]. Inert (nylon meshes, polyurethane foams, polypropylene rings) and organic (blocks, wood chips) support material have been used [93]. The main advantage of this technology is that can operate at organic loading between 69 and 142 times greater than anaerobic lagoons; but have clogging problems [93].

The most advance technology has been developed to improve operational problems of conventional technologies. On the one hand, AFBR systems (Anaerobic Fluidized Bed Reactor) are technologies studied mainly at laboratory scale. They use support material (clay, wood and PVC), which is suspended due to the recirculation of the flow [102]. These characteristics partially avoid clogging [103]. On the other hand, UASB technology has been applied at the laboratory and pilot level. These treatment units generate biomass granulated, which sediments (4 m/h) improving the cellular retention time [102]. The flow goes up, favoring the washing of biomass non-granulated the granules are dense, harboring multi-species and diverse microbial communities. However, the granulation processes require longer periods of formation (2–8 months). Other innovative technologies from UASB systems are the EGSB (*Expanded Granular Sludge Bed*). This last technology is hydraulically improved respect to UASB, because it operates at greater flow velocity (>4 m/h) than UASB [104]. In general, UASB reactors are a viable alternative, since they are considered high load systems, operating at organic loadings between 2 and 162 times higher than conventional systems (anaerobic lagoons, CSRT). The organic matter removal efficiencies reach ranges between 19 and 86% [93, 94, 103, 105]. In addition, they offer other operational advantages related to their volume (0.006–0.5 times less volume than lagoons and CSRT) and sludge production (granular from UASB vs. suspended from lagoons/CSRT) [104]. Moreover, UASB generates higher biogas production (0.28–4.05 m^3/m^3 d) than conventional systems (0.02–1.69 m^3/m^3 d), thanks to the fact that they operate at higher organic loading (1–8.1 kg COD/m^3 d) [69, 93].

Table 6 and **Figure 2** describe the operational characteristics of anaerobic technologies applied on swine slurries.

4.2 Nutrients and metals removal

Constructed wetlands are used as a cost-effective alternative for the nutrients removal within the livestock sector [70]. In Europe, there are around 60 livestock farms, which treat 17,000 tons/year using constructed wetlands [87]. In the United States, about 33% (~ 70 farms) of constructed wetlands are used within livestock sector, being mainly (~ 83%) surface flow constructed wetlands (SF-CW) [70].

Operationally, there is experience in the use of different types of constructed wetlands within swine sector. However, SF-CW are the most used technology, mainly to avoid clogging [106]. Generally, SF-CW are used after the anaerobic lagoon, operating at nutrient loading between 5 and 36 kg N/ha and between 1 and 6 kg P/ha. The nitrogen and phosphorous removal efficiencies obtained reach values from 50 and 90 to and 25–66%, respectively [106–109]. Moreover, horizontal subsurface flow constructed wetlands HSS-CW have been studied at laboratory and pilot scales, operating at nutrient loading between 69 and 252 kg N/ha d and

Technology	Loading kg (BOD_5^*, COD^, VS°) /m^3 d	Temperature °C	Efficiency % (BOD_5^*, COD^, VS°)	Biogas m^3 biogas/ kg (BOD_5^*, COD^, VS°)	% Methane	Reference
Lagoon	0.05–0.08°	24–32	26–79*	0.43–0.80°	86–95	[69, 93, 97, 98]
CSRT	0.41–2.04°	20–60	25–74*	0.19–0.83°	60–79	[93, 94, 103]
AF	3.44–11.34°	31–55	35–61^	0.03–0.29°	61–87	[93]
AFBR	1.1–6.6^	—	66–91^	0.17–0.53^	75–84	[103]
UASB	1.0–8.1^	20.7–35	19–86*	0.28–0.50^	54–87	[93, 94, 103, 105]

CSRT, continuous stirred reactor tank, AF, anaerobic filter, AFBR, anaerobic fluidized bed reactor, UASB, upflow anaerobic sludge blanket.
** =BOD5 units, ^ = COD units,° = VS units.*

Table 6.
Operational characteristics of anaerobic reactors within swine sector.

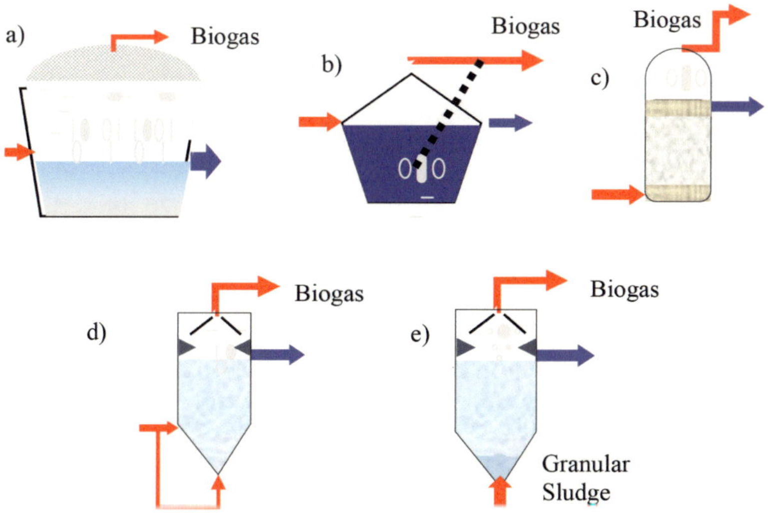

Figure 2.
Schemes of different types of anaerobic reactors used in the swine slurry treatment. (a) Lagoon, (b) CSRT, (c) AF, (d) AFBR, (e) UASB.

Species	Type	Prior Treatment	HRT (days)	Loading rate		Efficiency (%)		Reference
				kg N/ ha d	kg P/ ha d	N	P	
Tl, Sa	6 (HSS-lagoon-HS)	Anaerobic Lagoon	18	7–40	3.0–22.0	37–51^	13–31^	[109]
Sa, Sc, Sv, Je, Spa, Ta, Tl	4HSS	Anaerobic Lagoon Nitrifying unit	11–13	4.8–27.2 44–51	0.9–6.0 7–9	50–84^ 78–88^	25–38^ <10^	[107, 108, 106]
Ec	1HSS	Activated Sludge	27.5	69.0–262.0	15.0-47.0	10.0-24.0^	47.0-59.0^	[110]
Pc, Pa, Tl	Hybrid (2VS + 1HSS)	Anaerobic lagoon + Sand filter Recirculated treated purine (25–100%)	— —	214 —	30 —	50^ 54–67^	42^ 47–49^	[112]
Pa, Tl, Cp, Ca, Sl, Sm, Fu, Is, Ma, Ac, Se, Apa, Ls, Rl, S, P, Lm, Sta, Msp, Eca	Hybrid 8(HS+ HSS + VS)	Activated sludge	—	1.2–10	0.4–1	99.3^	99.6^	[113]
Gm, Ga, Gs, Mm, Poa	16HSS	Anaerobic digester	—	11.2–36.0	1.1–1.8	78–90*	56–66^	[106]
Cdp, Ap, Tl	2HSS	Filtration Tank	4.8	93.3	22.1	—	—	[111]
Ci, So, Pa	Híbrido (3 VS + 1 HSS)	Aeration lagoon	4–5	76.3	2.6	64^	61^	[114]

Acorus calamus: Ac; Alisma plantago-aquatica: Apa; Althernanthera philoxeroides: Ap; Canna indica: Ci; Carex pseudocyperus: Cp; Carex acutiformis: Ca; Cynodon dactylon Pers: Cdp; Elodea canadensis: Eca; Eichhornia crassipes: Ec; Filipendula ulmaria: Fu; Glyceria aquatica: Ga; Glyceria maxima: Gm; Iris pseudacorus: Ip; Juncus effusus: Je; Lemna minor: Lm; Lythrum salicaria: Ls; Mentha aquatica: Ma; Myriophyllum spicatum: Msp; Molinia maxima: Mm; Phragmites australis: Pa; Phragmites communis: Pc; Poa aquatic: Poa; Populus spp.: P; Ranunculus lingua: Rl; Salix spp.: S; Scirpus lacustris: Sl; Scirpus maritimus: Sm; Scirpus validus: Sv; Scirpus cyperinus: Sc; Schoenoplectus americanus: Sa; Sparganium erectum: Se; Sparganium americanum: Spa; Stratiotes aloides: Sta; Symphytum officinale: So; Typha angustifolia: Ta; Typha latifolia: Tl.
^ = *NH4+*

Table 7.
Operational characteristics of constructed wetlands within swine sector.

between 15 and 47 kg P/ha d. The nitrogen and phosphorus removal efficiencies vary from 10 to 24 and 47 to 59%, respectively [110, 111]. There are also experiences hybrid systems (SF/HSS/VSS) operating with plant species emergent and floating, which have achieved nitrogen and phosphorus removal efficiencies higher than 50 and 42%, respectively [112–114]. In general, constructed wetland systems will be operationally work with any previous technology. Thus, activated sludge has been used prior to constructed wetland increasing from 2 to 20 times the nutrients removal than anaerobic lagoons [109, 113]. The metals removal efficiencies have been reported in SF-CW, which have operated at rates from 0.09 to 0.25 kg Cu/ha and 0.58 to 1.58 kg Zn/ha d, obtaining removal efficiencies of up to 83 and 92%, respectively [78, 79, 111].

Currently, there are some innovations related to the constructed wetland treatment [106]. Likewise, the partial recirculation of pre-nitrified slurry has allowed to increase the nitrogen removal via denitrification up to 4 times, decreasing the ammonium volatilization [112, 115]. Other design concepts are based on the use of "*marsh-pond-marsh*" [109]. Constructed wetland technologies, could be improved operationally using intermittent hydraulic rate, which favors the oxygenation improving the nitrification [116].

Table 7 details the constructed wetlands operational characteristics of constructed wetlands used in the pig sector.

5. Future perspective and conclusions

Currently, the swine production should be looking to set the "new zoo technical order" with improvements in the life quality of the animals. Some reasons are given by environmental and health concerns given by the presence of emerging pollutants in meat and animal excreta. Indeed, swine meat has been reported as one of the sources of staphylococcus microbial resistance in humans [117]. Moreover, studies evidence the consumption of about 63,000 ton/year antibiotics in the livestock production (veterinary/promoters), being the main source of emerging pollutants in swine excreta/slurry [118]. Both water bodies and soil can be affected when these wastes are discharged or revaluated, since current treatment technologies are not designed to remove them. These two factors are further enhanced by the greenhouse gasses emissions responsibility from livestock production. Thus, swine production generates about 24 kg CO_2 eq/kg protein, which is mainly attributed to the mismanagement of their excreta/slurries [119]. In this last aspect, it is that the closing of the cycle the generation, treatment and revaluation of swine excreta fulfills a fundamental role. Studies report that reductions of up to 30% in greenhouse emissions could be achieved by a comprehensive management of resources (slurry, excreta, crops remains) between livestock and crop production [120]. Thus, livestock production through appropriate technology and management practices can be a source of nutrients for crops that provide food to animals. Agricultural production would support the energy generation inside farms by anaerobic. In the future, the livestock production could be supported from integral improvement from animal production to treatment and revaluation of wastes.

Author details

Cristina Alejandra Villamar* and Cristóbal Sardá
Departamento de Ingeniería en Obras Civiles, Facultad de Ingeniería, Universidad de Santiago de Chile, Santiago, Chile

*Address all correspondence to: cristina.villamar@usach.cl

IntechOpen

© 2019 The Author(s). Licensee IntechOpen. This chapter is distributed under the terms of the Creative Commons Attribution License (http://creativecommons.org/licenses/by/3.0), which permits unrestricted use, distribution, and reproduction in any medium, provided the original work is properly cited.

References

[1] Steinfeld H, Gerber P, Wassenaar T, Castel V, Rosales M, de Haan C. La larga sombra del ganado: problemas ambientales y opciones. Italia, Roma: División de Comunicación de la FAO; 2009. pp. 465

[2] Delgado C, Rosegrant H, Steinfeld H, Ehui S, Courbois C. Livestock to 2020. The Next Food Revolution. Food, Agriculture, and the Environment Discussion Paper 28. Washington D.C., United States: International Food Policy Research Institute; 1999. pp. 83

[3] FAOSTAT. Organización de las Naciones Unidas para la Alimentación y la Agricultura, Datos, Ganadería [Internet]. 2019. Available from: http://www.fao.org/faostat/es/#data/QA

[4] Froese C. Water usage and manure production rates in today's pig industry. Advances in Pork Production. 2003;**14**:218-223

[5] Solé F, Flotats X. Guía de técnicas de gestión ambiental de residuos agrarios. Catalunya, España: Centre UdL-IRTA; 2004

[6] Portejoie S, Dourmad J, Martinez J, Lebreton Y. Effect of lowering dietary crude protein on nitrogen excretion, manure composition and ammonia emission from fattening pigs. Livestock Production Science. 2004;**91**(1-2):45-55

[7] Galassi G, Colombini S, Malagutti L, Crovetto GM, Rapetti L. Effects of high fibre and low protein diets on performance, digestibility, nitrogen excretion and ammonia emission in the heavy pig. Animal Feed Science and Technology. 2010;**161**(3-4):140-148

[8] Kruse S, Traulsen I, Krieter J. Analysis of water, feed intake and performance of lactating sows. Livestock Science. 2011a;**135**(2-3):177-183

[9] Dourmad JY, Jondreville C. Impact of nutrition on nitrogen, phosphorus, Cu and Zn in pig manure and on emissions of ammonia and odours. Livestock Science. 2007;**112**(3):192-198

[10] Basset-Mens C, Van der Werf HMG. Scenario-based environmental assessment of farming systems: The case of pig production in France. Agriculture, Ecosystems and Environment. 2005;**105**(1-2):127-144

[11] Santomá G, Pontes M. Influencia del alojamiento sobre la nutrición de aves y cerdos. Barcelona, 16-17 octubre. XXII Curso de especialización FEDNA; 2006

[12] Dong GZ, Pluske JR. The low feed intake in newly weaned pigs: Problems and possible solutions. Asian-Australasian Journal of Animal Sciences. 2007;**20**(3):440-452

[13] Hernández F, Martínez S, López C, Megías MD, López M, Madrid J. Effect of dietary crude protein levels in a commercial range, on the nitrogen balance, ammonia emission and pollutant characteristics of slurry in fattening pigs. Animal: An International Journal of Animal Bioscience. 2011;**5**(8):1290-1298

[14] Dourmad JY, Guingand N, Latimier P, Séve B. Nitrogen and phosphorus consumption, utilisation and losses in pig production: France. Livestock Production Science. 1999;**58**:199-211

[15] Gonyou HW, Stricklin WR. Effects of floor area allowance and group size on the productivity of growing/finishing pigs. Journal of Animal Science. 1998;**76**:1326-1330

[16] Ward D, McKague K. Water Requirements of Livestock. Food and Rural Affairs, Ontario: Fact-sheet Ministry of Agriculture; 2007

[17] Jondreville C, Revy PS, Dourmad JY. Dietary menas to better control the environmental impact of copper and zinc by pigs from weaning to slaugter. Livestock Production Science. 2003;**84**(2):147-156

[18] Moral R, Perez-Murcia MD, Perez-Espinosa A, Moreno-Caselles J, Paredes C. Estimation of nutrient values of pig slurries in Southeast Spain using easily determined properties. Waste Management. 2005;**25**(7):719-725

[19] Scotford IM, Cumby TR, White RP, Carton OT, Lorenz F, Hatterman U, et al. Estimation of the nutrient value of agricultural slurries by measurement of physical and chemical properties. Journal of Agricultural Engineering Research. 1998;**71**:291-305

[20] Martínez-Suller L, Azzellino A, Provolo G. Analysis of livestock slurries from farms across northern Italy: Relationship between indicators and nutrient content. Biosystems Engineering. 2008;**99**(4):540-552

[21] Provolo G, Martínez-Suller L. In situ determination of slurry nutrient content by electrical conductivity. Bioresource Technology. 2007;**98**(17):3235-3242

[22] Suresh A, Choi HL. Estimation of nutrients and organic matter in Korean swine slurry using multiple regression analysis of physical and chemical properties. Bioresource Technology. 2011;**102**(19):8848-8859

[23] Ciampitti IA, García FO. Requerimientos nutricionales Absorción y extracción de macronutrientes y micronutrientes. Cereales, oleaginosos e industriales. Buenos Aires, Argentina; INPOFOS Archivos agronómico. 2007. pp. 13-15

[24] Moral R, Perez-Murcia MD, Perez-Espinosa A, Moreno-Caselles J, Paredes C, Rufete B. Salinity, organic content, micronutrients and heavy metals in pig slurries from South-Eastern Spain. Waste Management. 2008;**28**(2):367-371

[25] Boursier H, Béline F, Paul E. Piggery wastewater characterization for biological nitrogen removal process design. Bioresource Technology. 2005;**96**(3):351-358

[26] Marcato CE, Pinelli E, Pouech P, Winterton P, Guiresse M. Particle size and metal distributions in anaerobically digested pig slurry. Bioresource Technology. 2008;**99**(7):2340-2348

[27] De la Torre AI, Jimenéz JA, Carballo M, Fernandez C, Roset J, Muñoz MJ. Ecotoxicological evaluation of pig slurry. Chemosphere. 2000;**41**(10):1629-1635

[28] Villamar CA, Cañuta T, Belmonte M, Vidal G. Characterization of swine wastewater by toxicity identification evaluation methodology (TIE). Water, Air, and Soil Pollution. 2011;**223**:363-369

[29] Hoekstra NJ, Bosker T, Lantinga EA. Effects of cattle dung farms with different feeding strategies on germination and initial root growth of cress (*Lepidium sativum* L.). Agriculture, Ecosystems and Environment. 2002;**93**(1-3):189-196

[30] Li YS, Robin P, Cluzeau D, Bouché M, Qiu JP, Laplanche A, et al. Vermifiltration as a stage in reuse of swine wastewater: Monitoring methodology on an experimental farm. Ecological Engineering. 2008;**32**(4):301-309

[31] Mangas-Ramírez E, Sarma SSS, Nandini S. Combined effects of algal (*Chorella vulgaris*) density and ammonia concentration on the population dynamics of *Ceriodaphnia dubia* and *Monia macrocopa* (Cladocera). Ecotoxicology and Environmental Safety. 2002;**51**(3):216-222

[32] Dave G, Nilsson E. Increased reproductive toxicity of landfill leachate after degradation was caused by nitrate. Aquatic Toxicology. 2005;**73**(1):11-30

[33] Leung J, Kumar M, Glatz P, Kind K. Impacts of un-ionized ammonia in digested piggery effluent on reproductive performance and longevity of *Daphnia carinata* and *Monia australiensis*. Aquaculture. 2011;**310**(3-4):401-406

[34] Pretti C, Chiappe C, Baldetti I, Brunini S, Monni G, Intorre L. Acute toxicity of ionic liquids for three freshwater organisms: *Pseudokirchneriella subcapitata*, *Daphnia magna* and *Danio rerio*. Ecotoxicology and Environmental Safety. 2009;**72**(4):1170-1176

[35] Ip YK, Chew DJ, Randall DJ. Ammonia toxicity, tolerance, and excretion. Fish Physiology. 2001;**20**:109-148

[36] Kir M, Kumlu M, Eroldogan OT. Effects of temperature on acute toxicity of ammonia to *Penaeus semisulcatus* juveniles. Aquaculture. 2004;**241**(1):479-489

[37] Britto DT, Kronzucker HJ. NH_4^+ toxicity in higher plants: A critical review. Journal of Plant Physiology. 2002;**159**(6):567-584

[38] Van Der Eerden LJM. Toxicity of ammonia to plants. Agriculture and Environment. 1982;**7**(3-4):223-235

[39] Waller DL, Fisher SW, Dabrowska H. Prevention of zebra mussel infestation and dispersal during aquaculture activities. The Progressive Fish-Culturist. 1996;**58**(2):77-84

[40] Bernstein L. Effects of salinity and sodicity on plant growth. Annual Review of Phytopathology. 1975;**13**:295-312

[41] Martins J, Oliva Teles L, Vasconcelos V. Assays with *Daphnia magna* and *Danio rerio* as alert systems in. Aquatic Toxicology. 2007;**33**(3):414-425

[42] Shah FUR, Ahmad N, Masood KR, Peralta-Videla JR, Ahmad FuD. Heavy metal toxicity in plants. In: Plant Adaptation and Phytoremediation. New York, USA. Springer Publisher; 2010. pp. 71-97

[43] Lukkari T, Aatsinki M, Väisänen A, Haimi J. Toxicity of copper and zinc assessed with three different earthworm tests. Applied Soil Ecology. 2005;**30**(2):133-146

[44] Muyssen BTA, Janssen CR. Age and exposure duration as a factor influencing Cu and Zn toxicity toward *Daphnia magna*. Ecotoxicology and Environmental Safety. 2007;**68**(3):436-442

[45] Kristin MG, Todd VR. Effect of road salt application on seasonal chloride concentrations and toxicity in south-Central Indiana streams. Journal of Environmental Quality. 2010;**39**:1036-1042

[46] Blanes-Vidal V, Hansen MN, Adamsen APS, Feilberg A, Petersen SO, Jensen BB. Characterization of odor released during handling of swine slurry: Part I. relationship between odorants and perceived odor concentrations. Atmospheric Environment. 2009a;**43**(18):2997-3005

[47] Blanes-Vidal V, Hansen MN, Adamsen APS, Feilberg A, Petersen SO, Jensen BB. Characterization of odor released during handling of swine slurry: Part II. Effect of production type, storage and physicochemical characteristics of the slurry. Atmospheric Environment. 2009b;**43**(18):3006-3014

[48] Mackie RI, Stroot PG, Varel VH. Biochemical identification and biological origin of key odor components in livestock waste. Journal of Animal Science. 1998;**76**(5):1331-1342

[49] Schiffman SS, Bennett JL, Raymer JH. Quantification of odors and odorants from swine operations in North Carolina. Agricultural and Forest Meteorology. 2001;**108**(3):213-240

[50] Zhu J. A review of microbiology in swine manure odor control. Agriculture, Ecosystems and Environment. 2000;**78**(2):93-106

[51] Welsh FW, Schulte DD, Kroeker EJ, Lapp HM. The effect of anaerobic digestion upon swine manure odors. Canadian Agricultural Engineering. 1977;**19**(2):122-126

[52] Chae KJ, Jang A, Yim SK, Kim IS. The effects of digestion temperature and temperature shock on the biogas yields from the mesophilic anaerobic digestion of swine manure. Bioresource Technology. 2008;**99**(1):1-6

[53] Espinoza-Escalante FM, Pelayo-Ortíz C, Navarro-Corona J, González-García Y, Bories A, Gutiérrez-Pulido H. Anaerobic digestion of the vinasses from the fermentation of Agave tequilana weber to tequila: The effect of pH, temperature and hydraulic time on the production of hydrogen and methane. Biomass and Bioenergy. 2009;**33**(1):14-20

[54] Sheffield R, Thompson M, Dye B, Parker D. Evaluation of field-based odor assessment methods. Proceedings of the water environment federation. WEF/A&WMA Odors and Air emissions. 2004;**10**:870-879

[55] Pan L, Yang SX, DeBruyn J. Factor analysis of downwind odours from livestock farms. Biosystems Engineering. 2007;**96**(3):387-397

[56] Hanajima D, Kuroda K, Morishita K, Fujita J, Maeda K, Morioka K. Key odor components responsible for the impact on olfactory sense during swine feces composting. Bioresource Technology. 2010;**101**(7):2306-2310

[57] Nimmermark S. Influence of odour concentration and individual odour thresholds on the hedonic tone of odour from animal production. Biosystems Engineering. 2011;**108**(3):211-219

[58] Pomar C. Alimentar mejor a los cerdos para reducir el impacto medio ambiental. Jornadas técnicas: Factores que afectan la eficiencia productiva y la calidad en porcino. Vic, 1 de junio de 1999; 1999

[59] Jacela JY, DeRounchey JM, Tokach MD. Feed additives fro swine: Fact sheets – High dietary levels of copper and zinc for young pigs, and phytase. Journal Swine Health Production. 2010;**18**(2):87-91

[60] McEwen SA. Antibiotic use in animal agriculture: What have we learned and where are we going? Animal Biotechnology. 2006;**17**:239-250

[61] Choi F. Piggery Waste Management. Towards a Sustainable Future. London, United Kingdom: IWA Publishing; 2007. pp. 167

[62] Babot D, Martinez L, Teira MR. Gestión de subproductos y residuos porcinos. Mundo ganadero. 2001;**133**:34-36

[63] Martínez-Almela J, Barrera JM. SELCO-Ecopurin pig slurry treatment system. Bioresource Technology. 2005;**96**(2):223-228

[64] Aust M-O, Thiele-Bruhn S, Eckhardt K-U, Leinweber P. Composition

of organic matter in particle size fractionated pig slurry. Bioresource Technology. 2009;**100**(23):5736-5743

[65] Bernet N, Béline F. Challenger and innovations on biological treatment of livestock effluents. Bioresource Technology. 2009;**100**(22):5431-5436

[66] Vanotti MB, Szogi AA, Millner PD, Loughrin JH. Development of a second-generation environmentally superior technology for treatment of swine manure in the USA. Bioresource Technology. 2009;**100**(22): 5406-5416

[67] Safley LM Jr, Wasterman PW. Biogas production from anaerobic lagoons. Biological Wastes. 1988;**23**:181-193

[68] Knight RL, Payne VWE Jr, Borer RE, Clarke RA Jr, Pries JH. Constructed wetlands for livestock wastewater management. Ecological Engineering. 2000;**15**:41-55

[69] Vanotti MB, Szogi AA, Hunt PG, Millner PD, Humenik FJ. Development of environmentally superior treatment system to replace anaerobic swine lagoons in the USA. Bioresource Technology. 2007;**98**(17):3184-3194

[70] Belmonte M, Vázquez-Padín JR, Figueroa M, Campos JL, Méndez R, Vidal G. Denitrifying activity via nitrite and N_2O production using acetate and swine wastewater. Process Biochemistry. 2012;**47**(7):1202-1206

[71] Bonmatí A, Flotats X. Air stripping of ammonia from pig slurry: Characterization and feasibility as a pre- or post-treatment to mesophilic anaerobic digestion. Waste Management. 2002;**23**(3):261-272

[72] Bonmatí A, Flotats X. Pig slurry concentration by vacuum evaporation: Influence of previous mesophilic anaerobic digestion process. Air and Waste Management. 2003;**53**(1):21-31

[73] Neubauer ME, Plaza de los Reyes C, Pozo G, Villamar CA, Vidal G. Growth and nutrient uptake by *Schoenoplectus californicus* (C.A. Méyer) Sójak in a constructed wetland fed with swine slurry. Journal of Soil Science and Plant Nutrition. 2012;**12**(3):421-430

[74] Plaza de los Reyes C, Villamar CA, Neubauer M, Pozo G, Vidal G. Behavior of *Typha angustifolia* L. in a free water surface constructed wetland for the treatment of swine wastewater. Journal of Environment Science and Health, Part A. 2013;**48**(10):1216-1224

[75] Villamar CA, Rivera D, Neubauer ME, Vidal G. Nitrogen and phosphorus dynamics in a constructed wetland fed with treated swine slurry from an anaerobic lagoon. Journal of Environmental Science and Health: Part: A. 2015;**50**(1):60-71

[76] Yeh TY, Chou CC, Pan CT. Heavy metal removal within pilot-scale constructed wetlands receiving river water contaminated by confined swine operations. Desalination. 2009;**249**(1):368-373

[77] Villamar CA, Neubauer ME, Vidal G. Distribution and availability of copper and zinc in a constructed wetland fed with treated swine slurry from an anaerobic lagoon. Wetlands. 2014;**34**(3):583-591

[78] Sánchez M, González JL. The fertilizer value of pig slurry. I. Values depending on the type of operation. Bioresource Technology. 2005;**96**(10):1117-1123

[79] Ndayegamiye A, Coté D. Effect of long-term pig slurry and solid cattle manure application on soil chemical and biological properties. Canadian Journal of Soil Science. 1989;**69**:39-47

[80] Wang H, Magesan G, Bolan N. An overview of the environmental effects of land application of farm effluents. New Zealand Journal of Agricultural Research. 2004;**47**(4):389-403

[81] Teira-Esmatges MR, Flotats X. A method for livestock waste management planning in NE Spain. Waste Management. 2003;**23**(10):917-932

[82] Bastian R, Murray D. Guidelines for Water Reuse, EPA/600/R-12/618. Washington, DC, USA: US EPA Office of Research and Development; 2012. p. 643

[83] Provolo G. Manure management practices in Lombardy (Italy). Bioresource Technology. 2005;**96**(2):145-152

[84] Villamar CA, Vera I, Rivera D, de la Hoz F. Reuse and recycling of livestock and municipal wastewater in Chilean agriculture: A preliminary assessment. Water. 2018;**10**(6):817-833

[85] Henning F, Flotats X, Bonmati A, Palatsi J, Magri A, Schelde KM. Inventory of manure processing activities in Europe. Technical Report No. I concerning "Manure Processing Activities in Europe" to the European Commission, Directorate-General Environment; 2011. pp. 138

[86] Angelidaki I, Ellegaard L. Codigestion of manure and organic wastes in centralized biogas plants. Applied Biochemistry and Biotechnology. 2003;**109**:95-105

[87] Lehtomäki A, Huttunen S, Rintala JA. Laboratory investigations on co-digestion of energy crops and crop residues with cow manure for methane production: Effect of crop to manure ratio. Resources, Conservation and Recycling. 2007;**51**:591-609

[88] Al Seadi T. Danish Centralised Biogas Plants-Plants Descriptions. Esbjerg, Demarck: Bioenergy Department, University of Southern Denmark; 2000. pp. 27

[89] Weiland P. Anaerobic waste digestion in Germany –status and recent developments. Biodegradation. 2000;**11**(6):415-421

[90] Villamar CA, Rivera D, Aguayo M. Anaerobic co-digestion plants for the revaluation of agricultural waste: Sustainable location sites from a GIS analysis. Waste Management & Research. 2016;**34**(4):316-326

[91] Chynoweth DP, Wilkie AC, Owens JM. Anaerobic processing of piggery wastes: a review. ASAE Annual International Meeting. Orlando, Florida, July 11-16; 1998

[92] Lusk P. Methane Recovery from Animal Manures the Current Opportunities Casebook. 3rd ed. Washington D.C., United States: NREL/SR-25145.Golden, CO: National Renewable Energy Laboratory. Work performed by Resource Development Associates; 1998. pp. 174

[93] Wu B, Chen Z. An integrated physical and biological model for anaerobic lagoons. Bioresource Technology. 2010;**102**(9):5032-5038

[94] Barker JC. Lagoon design and management for livestock waste treatment and storage. North Carolina Cooperative Extension Service, Pub. No. EBAE, 103-83; 1996

[95] Hart SA, Turner ME. Lagoons for livestock manure. Water Environment Federation. 1965;**37**(11):1578-1596

[96] McLaughlin MR, Brooks JP, Adeli A. Temporal flux and spatial dynamics of nutrients, fecal indicators, and zoonotic pathogens in anaerobic swine manure lagoon water. Water Research. 2012;**46**(16):4949-4960

[97] Lovanh N, Loughrin JH, Cook K, Rothrock M, Sistani K. The effect of stratification and seasonal variability on the profile of an anaerobic swine waste treatment lagoon. Bioresource Technology. 2009;**100**(15):3706-3712

[98] Hill VR. Prospects for pathogen reductions in livestock wastewater: A review. Critical Reviews in Environmental Science and Technology. 2003;**33**(2):187-235

[99] Lim T-T, Heber AJ, Ni J-Q, Shao P. Atmospheric pollutans and trace gases: Odor and gas release from anaerobic treatment lagoons for swine manure. Journal of Environmental Quality 2003;**32**:406-416

[100] Liu X, Yan Z, Yue Z-B. Biofuels and bioenergy/biogas. In: Moo-Young M, editor. Comprehensive Biotechnology. Second ed. Vol. V3. 2011. pp. 99-144

[101] Montalvo S, Gerrero L. Tratamiento anaerobio de residuos. Producción de biogás. Valparaíso, Chile: Universidad Técnica Santa María; 2003. pp. 413

[102] Seghezzo L, Zeeman G, van Lier JB, Hamelers HVM, Lettinga G. A review: The anaerobic treatment of sewage in UASB and EGSB reactors. Bioresource Technology. 1998;**65**(3):175-190

[103] Rodríguez DC, Belmonte M, Peñuela G, Campos JL, Vidal G. Behaviour of molecular weight distribution for the liquid fraction of pig slurry treated by anaerobic digestion. Environmental Technology. 2011;**32**(4):419-425

[104] Harrington C, Scholz M. Assessment of pre-digested piggery wastewater treatment operations with surface flow integrated constructed wetland systems. Bioresource Technology. 2010;**101**(20):7713-7723

[105] Stone KC, Hunt PG, Szögi AA, Humenik FJ, Rice JM. Constructed wetlsn design and performance for swine lagoon wasteeater treatment. American Society of Agricultural Engineers. 2002;**45**(3):723-730

[106] Hunt PG, Szögi AA, Humenick FJ, Rice JM, Matheny TA, Stone KC. Constructed wetlands for treatment of swine wastewater from an anaerobic lagoon. Soil science Society of America Journal. American Society of Agricultural Engineers. 2002;**45**(3):639-647

[107] Poach ME, Hunt PG, Reddy GB, Stone KC, Johnson MH, Grubbs A. Swine wastewater treatment by marsch-pond-marsh constructed wetlands under varyring nitrogen loads. Ecological Engineering. 2004;**23**(3):165-175

[108] Lee C-Y, Lee C-C, Lee F-Y, Tseng S-K, Liao C-J. Performance of subsurface flow constructed wetland taking pretreated swine effluent under heavy loads. Bioresource Technology. 2004;**92**:173-179

[109] Matos AT, Freitas S, Martinez MA, tótola MR, Azevedo AA. Tifton grass yield on constructed wetland used for swine wastewater treatment. Revista Brasileira de Engenharia Agrícola e Ambiental. 2010;**14**(5):510-516

[110] Lian-sheng H, Hong-liang L, Bei-dou X, Ying-bo Z. Enhancing treatment effiency of swine wastewater by effluent recirculation in vertical-flow constructed wetland. Journal of Environemtnal Sciences. 2006;**18**(2):221-226

[111] Meers E, Tack FMG, Tolpe I, Michels E. Application of a full-scale constructed wetlsn for tertiary treatment of piggery manure: mo itoring results. Water, Air, and Soil Pollution. 2008;**193**(1):15-24

[112] Borin M, Politeo M, De Stefani G. Performance of a hybrid constructed wetland treating piggery wastewater. Ecological Engineering. 2013;**51**:229-236

[113] Hunt PG, Stone KC, Matheny TA, Poach ME, Vanotti MB, Ducey TF. Denitrification of nitrified and non-nitrified swine lagoon wastewater in the suspended sludge layer of treatment wetlands. Ecological Engineering. 2009;**35**(10):1514-1522

[114] Poach ME, Hunt PG, Reddy GB, Stone KC, Johnson MH, Grubbs A. Effect of intermittent drainage on swine wastewater treatment by marsch-pond-marsh constructed wetlands. Ecological Engineering. 2007;**30**(1):43-50

[115] Simeoni D, Rizzotti L, Cocconcelli P, Gazzola S, Dellaglio F, Torriani S. Antibiotic resistance genes and identification of staphylococci collected from the production chain of swine meat commodities. Food Microbiology. 2008;**25**(1):196-201

[116] Tasho RP, Cho JY. Veterinary antibiotics in animal waste, its distribution in soil and uptake by plants: A review. Science of the Total Environment. 2016;**563**:366-376

[117] Herrero M, Havlík P, Valin H, Notenbaert A, Rufino MC, Thornton PK, et al. Biomass use, production, feed efficiencies, and greenhouse gas emissions from global livestock systems. Proceedings of the National Academy of Sciences. 2013;**110**(52):20888-20893

[118] Adegbeye MJ, Reddy PRK, Obaisi Ai, Elghandour MMMY, Oyebamiji KJ, Salem AZM, et al. Sustainable agriculture options or production, nutritional mitigation o greenhouse gasses and prollution, and nutrient recycling in emergingn and transitional nations – An review. Journal of Cleaner Production. 2019 (In press)

Chapter 6

Livestock Feeds and Feeding in Semi-Arid Areas of Southern Africa

Clarice Princess Mudzengi, Everson Dahwa and Clayton Simbarashe Kapembeza

Abstract

Livestock production is the major source of rural livelihoods in semi-arid regions of Southern Africa. However, nutrition is the major limiting factor of livestock production in these areas characterised by declines in rangeland productivity due to the increases in drought frequency, deliberate overstocking by farmers, and climate change and variability. For instance, the grazing resource is strongly influenced by seasonality of rainfall. Poor-quality cereal crop residues are the main dry season supplementary feed source, yet the predominant crops such as sorghum and maize are deficient in protein and other essential nutrients. Additionally, although conventional supplements, fodder crops and agro by-products are an alternative dry season supplementary feed source; they are costly and not readily available. They are also mostly based on staple food crops such as maize, creating competition in use between humans and livestock. Therefore, indigenous browse species remain a significant source of abundant and persistent animal feeds. Other innovations with the potential to improve feed availability include straw ammoniation and silages, veld reinforcement and rehabilitation, and strategic destocking. However, they are not readily adopted by farmers. There is thus a need to promote technologies that improve livestock feeds and feeding for sustainable livelihoods.

Keywords: livestock production, nutrition, sustainable rural livelihoods

1. Introduction

In the semi-arid areas of Southern Africa, livestock production underpins the socio-economic and political lives of the people. Meat and milk from livestock are important dietary protein sources. Livestock production also creates employment opportunities and provides household income. Furthermore, it promotes viability of small-scale cropping systems through provision of draught power and organic manure. Cattle, in specific, are socioculturally important as a measure of wealth. Actually, Zimbabwean small-scale farmers generally own 89% of the national cattle herd, with the livestock sector contributing 35% of the agricultural gross domestic product [1]. However, constraints to the increasing livestock productivity in semi-arid areas include water and feed shortages, diseases, and lack of research and markets. Of these, nutrition is the major factor in extensive livestock production systems, contributing more than 75% of the total variable costs of production [2, 3]. Thus, a comprehensive inventory of animal feeds and feeding systems in semi-arid areas will inform sustainable livestock production.

Rangeland productivity, i.e. the amount of available grazing and browse per square area per unit time, is a proxy indicator of sustainability of livestock-based rural livelihoods. In recent years, climate change and variability, among other factors, has resulted in the declines in the quality and quantity of the rangelands in semi-arid areas such as the South East Lowveld (SEL) of Zimbabwe [4–7]. Additionally, in these areas, while Transfrontier Conservation Areas (TFCAs) have been established mainly to facilitate sustainable livelihoods, global biodiversity conservation, regional peace, and sustainable socio-economic development of African communities through the cooperation at local and international levels [8], they are also likely to increase interaction between wildlife, livestock, and humans with adverse consequences. For instance, due to increased human and livestock populations in surrounding agricultural areas, cattle are likely to encroach more into wildlife areas in search of feed [9]. Therefore, a deeper comprehension of animal feeds and feeding will improve livestock production and consequently transform rural livelihoods.

Innovations in livestock husbandry are the activities and processes associated with the generation, production, dissemination, adaptation, and use of existing or new technical, institutional, and organisational knowledge [10, 11]. Although there are different innovations in livestock feeds and feeding, most of them have not been adopted by farmers [12]. For instance, [12] showed that discontinuance of urea treatment of maize stover for livestock supplementation was attributed to high labour requirements of preparing the stover, lack of monitoring by extension services, and inaccessibility of urea fertiliser. It is thus important for the policy to consider such factors as the economic environment, availability of local material, and social and human capital when promoting livestock production systems. In this chapter, we explore and explain different livestock feeds and feeding strategies that are mostly adopted in semi-arid areas. We also recommend other alternatives that have a potential of adaption for increased livestock production.

2. Materials and methods

2.1 Study site

The study was carried out in the semi-arid South East Lowveld (SEL) of Zimbabwe. The area is found at an altitude of 300–600 m above mean sea level [11]. It experiences mean maximum and minimum temperatures of 21.8°C in October and 13.3°C in June, respectively, and mean annual rainfall of 300–600 mm between November and March and is characterised by high interannual variability (coefficient of variation ≈ 4045%) [13]. The major soil types are basalt-derived vertisols. Other soil types include eutric fluvisols, leptosols, and chromic luvisols [14]. The two main land uses in the area are agricultural production in the communal areas and wildlife conservation in Gonarezhou National Park and Malipati Safari Area, both of which form part of the Great Limpopo TFCA that contains a wide range of wildlife species such as *Loxodonta africana* (the elephant), *Giraffa camelopardalis* (giraffe), and *Syncerus caffer* (African buffalo). A communal land is a land category characterised by collective or community land ownership [15]. Livestock production is the major source of livelihoods in the communal area, while small grains and maize are also commonly grown.

2.2 Data collection

Individual structured questionnaires were administered to 150 respondents randomly selected, representing approximately 12% of the total households in the study area. The questionnaire was designed to capture socio-demography and livestock production characteristics, specifically feed resources and farmer innovations in livestock feeding and management. The questionnaire was pretested before final administration. We also conducted two focus group discussions (FGD) with seven key informants each, representative of the pastoral, agro-pastoral, and crop-livestock production systems in the area. For the woody species, we carried out veld assessment. Using the point-centred quarter method [16], we established 53 30 m × 30 m plots at each sampling point along 9 transects randomly established, measuring between 10 and 15 km each. The plant species were identified with the help of the locals in addition to using field identification guides [17–19]. Canopy structure, tree height, growth habit, leaf, bark, and other tree structures were used to differentiate closely related trees. Trees rooted within the plot, or along plot margins with at least half of the rooted system inside the plot, were considered [20]. We also recorded altitude and location of each individual tree using a Global Positioning System (GPS) Unit. Samples of the species not identified in the field, as well as all the other species, were collected for verification at the National Herbarium in Harare, Zimbabwe.

3. Results and discussion

3.1 Feeds and feeding resources in the SEL

3.1.1 Grazing resource

The veld of the SEL is described as "*Aristida-Dactyloctenium-Eragrostis* other species grassveld". It has a carrying capacity of 0.084–0.14 tropical livestock units per hectare [6]. The grazing period ranges from November/December to April/May. The veld remains nutritious and palatable for livestock across seasons. The herbaceous layer is dominated by *Aristida adscensionis* L., *Dactyloctenium giganteum* B.S. Fisher & Schweick., *Eragrostis viscosa* [Retz.] Trin., *Chloris virgata* Sw., and on deeper soils with more moisture, *Urochloa* spp., *Panicum* spp., *Cenchrus ciliaris* L., and *Digitaria* spp. [21]. On well-managed grazing systems, cattle exhibit annual live weight gains of 15 kg/ha. However, herbaceous species structure and composition are strongly influenced by seasonality of rainfall. For instance, the biomass disappears rapidly in drought years or when the start of the rainy season is delayed (**Figure 1**) [6]. Therefore, there is a need for supplementary feeding, especially during this period of scarcity of the grazing resource.

3.1.2 Crop residues

In the SEL, poor-quality cereal crop residues (less than 4% crude protein) form the bulk of livestock supplementary feed in the dry season, which normally extends from May/June to October/November. The predominant crops are sorghum, millet, and maize. However, they are deficient in essential nutrients such as protein, phosphorus, calcium, and, to some extent, energy [22]. Such supplements have low feed intake resulting from low degradability and low digestibility. Therefore, they do not provide for optimum microbial growth in the rumen. As a result, animals raised on these low nutritive feeds exhibit poor condition and reduced reproductive performance [23]. Crop residues are managed in many ways

Figure 1.
The grazing resource during dry seasons in the South East Lowveld of Zimbabwe.

Figure 2.
Maize stover forms part of the bulk of cereal stover supplements in the dry season.

for livestock feeding. Cereal stovers are either grazed in situ or stored in stacks for supplementation during the dry season (**Figure 2**). During prolonged dry seasons, the first preference is for maintenance of productive animals such as lactating cows or the sick. Haulms from leguminous crops such as cowpea and groundnuts are also used in stall-feeding. Despite being of higher nutritive value than cereal stovers, they have limited availability as leguminous crops are not commonly cultivated at large scale. Recently, conservation agriculture has presented conflict of interest in utilisation of crop residues. Conservation agriculture is a farming method that utilises crop residues to retain moisture and enrich the soil [24]. Increased adoption of conservation agriculture creates limitations in the availability of crop residues for livestock feeding.

3.1.3 Browse trees

Indigenous browse species are an important source of animal feed in livestock-based rural livelihoods of semi-arid areas (**Figure 3**) [25, 26]. The natural vegetation of the SEL is predominantly *Colophospermum mopane* [J.Kirk ex Benth.], J. Léonard woodlands found in association with *Kirkia acuminate* Oliv., *Dalbergia melanoxylon* Guill. & Perr, *Adansonia digitata* L., *Combretum* spp., *Acacia* spp., and *Commiphora* spp. In addition, recently, a shrubby legume called *Neorautanenia brachypus* [Harms] C.A.Sm. was discovered as a medicinal feed that helps livestock to survive drought [6]. Other browse species of the SEL are presented in **Table 1**. Most indigenous browse species remain abundant, evergreen, and relatively high in protein, metabolisable energy, vitamins, and minerals across seasons [27]. Unlike herbaceous species, browse species are less susceptible to climatic fluctuations, with crude protein (CP) levels of approximately 10% even in the dry season [28]. However, early and increased dependence on browse by livestock in semi-arid areas of the

Scientific name	Vernacular/English name
Acacia albida	Shokoshoko/winter thorn
Acacia karroo	Muunga/sweet thorn
Acacia tortilis	Sesani/umsasane/umbrella thorn
Acacia xanthophloea	Kelenga/fever tree
Adansonia digitata	Mabuwu/baobab/muwu
Aloe cameronii	Mhangani/aloe
Berchemia discolour	Munyii/bird plum
Boscia albitrunca	Shukutsu/shepherd's tree
Brachystegia spiciformis	Musasa
Cassia abbreviata	Murumanyama/long-tail cassia
Cissus quadrangularis	Chiololo/chiololoti/muvengahonye
Colophospermum mopane	Mopane/xanatsi/turpentine tree
Combretum apiculatum	Chikukutsi/red bushwillow
Combretum imberbe	Mutsviri/mondo/monzo/leadwood
Dichrostachys cinerea	Mupangara/ndenge/sickle bush
Diospyros mespiliformis	Musuma/tithoma/jackalberry
Ficus sycomorus	Muonde/mikuwa/sycamore fig
Hippocratea crenata	Sengeti/valley paddle pod
Hyphaene petersiana	Makwangwala/Ilala/real fan palm
Julbernadia globiflora	Mutondo
Kigelia africana	Pfungu/mumvewa/sausage tree
Lonchocarpus capassa	Mupanda/umchitamuzi/rain tree
Mimusops zeyheri	Hlatsva/Chechete/red milkwood
Neorautanenia brachypus	Zhombwe
Phragmites mauritianus	Shanga/reed grass
Salvadora persica	Dhungulu pokwe/mustard tree
Sclerocarya birrea	Mupfura/marula/mufura
Xanthocercis zambesiaca	Muhlaru/Musharo/Nyala berry

Adapted from Mudzengi et al. [36].

Table 1.
List of indigenous browse trees in the SEL.

SEL during the dry season limits their availability in the rest of the season [29]. Fresh leaves of species such as *C. mopane*, for instance, are high in tannins and lignin [30, 31]. Additionally, indigenous browse species normally attract multiple uses at the livestock-wildlife interface with the more visible, more dominant, and more frequent browse species having more uses than less apparent plants [32]. They are used as sources of firewood, timber, fruits, edible roots, bark and leaves, and human and ethnoveterinary medicines [33–36]. Competitive use increases vulnerability to overutilisation, unsustainable harvesting, and mismanagement.

3.1.4 Fodder crops

Fodder refers to any plants grown specifically as animal feed. They include a variety of pasture grasses like *Panicum maximum*, *Cenchrus ciliaris*, and *Chloris gayana*; pasture legumes such as *Vigna unguiculata*, *Dolichos lablab*, and *Macroptilium atropurpureum*; and fodder trees such as *Leucaena leucocephala*, *Acacia angustissima*, and *Calliandra calothyrsus*. However, most of them do not thrive in semi-arid areas such as the SEL due to high temperatures and low precipitation. Low adoption of fodder crop production is also attributed to lack of extension for farmer training, shortage of labour due to overlapping of the farming calendar with the main crop, high cost and unavailability of seed, and land scarcity. It is therefore important for farmers to maximise production of those species adaptable to their climatic conditions.

3.1.5 Conventional supplements, food industry, and agro-industrial by-products

There are different food industry by-products and agricultural wastes that are alternative dry season livestock feed supplements. These can be of animal and plant origin or of the fermentation industry. Animal by-products include blood, bones, meat and bone offals, fat, intestine and rumen contents, whey, tannery by-products, and poultry manure [1]. By-products of plant origin consist those of the milling industry (e.g. bran, waste flour), oil industry (e.g. soya bean and sunflower cakes), sugar industry (molasses), and citrus and horticulture waste. The fermentation industry produces grain, molasses, and brewer's waste, among a large array of other by-products. By-products of plant origin are the commonly used. For instance, in the SEL, molasses is readily available as the main sugarcane processing factories in Zimbabwe are located in that area. However, high cost of transportation

Figure 3.
Browse trees provide feed during the dry season when both the grazing resource and cereal stovers become limiting.

makes the product often quite expensive. Additionally, those of plant origin are also mostly based on staple food crops such as maize and soya bean, creating competition in use between humans and livestock.

3.2 Recommended innovations in livestock feeds and feeding

3.2.1 Straw ammoniation

Straw ammoniation is the process of addition of urea, liquid ammonia, or ammonium bicarbonate to poor-quality cereal crop stovers in order to improve their palatability, nutritive value, and digestibility. Of these three, urea is the most readily available and easiest to handle ammonia source. Nevertheless, in the SEL, as in most rural areas, urea treatment still is not a commonly used method improvement of the feeding value of cereal stovers due to lack of expertise in carrying out the procedure, as well as unaffordability for most rural resource-poor farmers. Additionally, if not done properly, urea-treated straw can be toxic to animals and cause air pollution.

3.2.2 Silages

Silage is forage produced from the fermentation process of chopped fresh green material under anaerobic conditions. These materials include fodder or forage grasses. Ensiling maize has been shown to improve feed digestibility and reduce methane gas production by 30% compared to feeding dry maize [37]. However, despite silages being advantageous in areas of water shortages, as well as reducing tannins due to the heat produced during the incubation period, silage production is not common among farmers.

3.2.3 Other strategies

Other potential technologies include intercropping cereals with ley (dual-purpose) legumes [38]. Ley legumes provide protein-rich fodder, improve the productivity of cereal crops by increasing the amount of nitrogen available for uptake, as well as offer a possible lower-cost alternative to nitrogen fertilisers [39, 40]. In addition to cut and carry systems for feeding fresh plant material, leaf meals can also be produced by drying harvested leaf material under shed. The commonly grown multipurpose trees include *Leucaena leucocephala*, *Calliandra calothyrsus*, and *Gliricidia sepium*. The leaf meal can then be incorporated in home-made livestock rations. Cutting and drying herbage from forage and multipurpose trees can also increase feed availability. Multipurpose trees can be grown in alleys as live boundaries, home gardens, and contour ridges and in woodlots.

3.3 Rangeland management

3.3.1 Principles of rangeland management

There are generally four fundamental principles of rangeland management which are important in order to improve condition and stability of the veld and consequently increase feed. They are rest, removal of top hamper, period of stay, and stocking density. Rest facilitates replenishment of plant growth reserves and also sets seed after defoliation. Top hamper is dead plant material accumulation which causes shading out of new shoots as they develop. It represents a nutrient bottleneck by preventing plant material from recycling back into the soil. It should therefore be removed. Another principle of rangeland management is control of period of utilisation by animals. This is important as too long periods result in overgrazing, while too short periods cause underutilisation which leads to top hamper and consequently reduced plant vigour. Stocking density refers to the number of animals that is kept on a given unit of area [41]. This has a direct relationship to the carrying capacity of the range. For instance, understocking causes selective grazing, which depletes palatable grass species. On the other hand, overstocking may degrade the range. For high stocks of reserve biomass, and for farmers with a relatively low degree of risk aversion, an "opportunistic" strategy is optimal, which matches the stocking rate with the available forage in every year [42]. On the other hand, the "resting in rainy years" grazing management strategies are recommended in which a lower stocking rate is applied in years in which current rainfall exceeds some threshold, and in years with current rainfall below this threshold, full stocking is optimal [39, 42–44].

3.3.2 Veld reinforcement

Veld reinforcement is the introduction, to the rangeland, of new grass or legume species in order to improve both the quantity and quality of the natural vegetation. Legume forages such as *Desmodium uncinatum*, *Macroptilium atropurpureum*, *Stylosanthes guianensis*, and *Cassia rotundifolia* can be used, while grass species including *Cynodon nlemfuensis*, *Paspalum notatum*, and *Panicum maximum* are also good for veld reinforcement.

3.3.3 Rangeland fertilisation

Rangeland fertilisation is the application of fertilisers such as ammonium nitrates on the rangelands in order to increase the quality and quantity of forage. However, this method is not highly recommended as fertilisers are expensive and at times not readily available.

3.3.4 Control of undesirable plants

Undesirable plants are not readily utilised by animals and may cause rangeland degradation. They include invasive species (e.g. *Dichrostachys cinerea)* and poisonous species (e.g. *Lantana camara* and *Solanum incanum*). They may be removed by stumping, ring barking, and application of chemicals such as arboricides, using hot prescribed fires or mechanical means like bulldozers, motorised saws, and brush cutters.

3.3.5 Range rehabilitation

Range rehabilitation is the restoration of the veld using such methods as gulley filling and planting grass lines. However, it is more feasible at small scale. Both communal and private enclosures have also been successfully used to rehabilitate rangelands [45, 46].

3.3.6 Strategic destocking

In the SEL, deterioration of rangeland productivity during prolonged dry season characteristic of the area is worsened by deliberate increases in cattle numbers by farmers who use the high cattle numbers as a hedge against losses during drought [6]. Therefore, it is recommended that farmers should adopt strategic destocking programmes that promote fattening of animals during periods of feed abundance and disposal while they are in good enough body condition to fetch high prices.

4. Conclusions

Livestock production in semi-arid areas is hampered by shortages of feed, especially in the dry season when the grazing resource becomes limiting. During this time, browse species play an important role as the most abundant and nutritious feed. Although crop residues are also a likely supplementary feed, they are of poor nutritive value. Potential technologies to improve such feedstuffs include urea treatment and ensilage. However, they are also not readily adopted by farmers due to lack of knowledge among other factors. It is therefore important to promote such innovations with the view to improve livestock production and hence rural livelihoods.

Acknowledgements

The authors would like to express great appreciation to Malipati traditional leadership in the SEL for the permission to conduct questionnaires and focus group discussions in their area, as well as collect test material.

Conflict of interest

The authors declare no conflict of interest.

Author details

Clarice Princess Mudzengi[1]*, Everson Dahwa[1] and Clayton Simbarashe Kapembeza[2]

1 Garry Magadzire School of Agriculture and Natural Sciences, Great Zimbabwe University, Masvingo, Zimbabwe

2 Department of Research and Specialist Services, Division of Livestock Research, Grasslands Research Institute, Marondera, Zimbabwe

*Address all correspondence to: clarice.mudzengi@gmail.com

IntechOpen

© 2020 The Author(s). Licensee IntechOpen. This chapter is distributed under the terms of the Creative Commons Attribution License (http://creativecommons.org/licenses/by/3.0), which permits unrestricted use, distribution, and reproduction in any medium, provided the original work is properly cited.

References

[1] Food and Agriculture Organisation [FAO]. Livestock Sector Briefs (Mozambique, Namibia, Zimbabwe). Rome: Livestock Information, Sector Analysis and Policy Branch (AGAL); 2005

[2] Matope A, Zindove TJ, Dhliwayo M, Chimonyo M. Mitigating the effects of drought on cattle production in communal rangelands of Zimbabwe. Tropical Animal Health and Production. 2019:1-10. DOI: 10.1007/s11250-019-02020-y

[3] Connolly L, Kinsella A, Quinlan G, Moran B. National Farm Survey. Athenry, Republic of Ireland: Teagasc; 2010

[4] Cumming DHM. Wildlife, livestock and food security in the south east lowveld of Zimbabwe. In: Osofsky SA et al, editors. Conservation and Development Interventions at the Wildlife/Livestock Interface: Implications for Wildlife, Livestock and Human Health, IUCN Occasional Paper No 3. Gland, Switzerland and Cambridge, United Kingdom; 2005. 41-46 pp

[5] Intergovernmental Panel on Climate Change (IPCC). Impacts, Adaptation and Vulnerability, Contribution of Working Group II to the Fourth Assessment Report of the Intergovernmental Panel on Climate Change (IPCC). Cambridge, United Kingdom; 2007

[6] Murungweni C, Andersson JA, van Wijk MT, Gwitira I, Giller KE. Zhombwe (*Neorautanenia brachypus* (Harms) C.A.Sm.)—A recent discovery for mitigating effects of drought on livestock in semi-arid areas of Southern Africa. Ethnobotany Research and Applications. 2012;**10**:199-212

[7] Descheemaeker K, Zijlstra M, Masikati P, Crespo O, Homann-Kee Tui S. Effects of climate change and adaptation on the livestock component of mixed farming systems: A modelling study from semi-arid Zimbabwe. Agricultural Systems. 2018;**159**:282-295

[8] de Garine-Wichatitsky M, Caron A, Gomo C, Foggin C, Dutlow K, Pfukenyi D, et al. Bovine tuberculosis in buffaloes, Southern Africa. Emerging Infectious Diseases. 2010;**16**(5):884-885

[9] Murwira A, de Garine-Wichatitsky M, Zengeya F, Poshiwa X, Matema S, Caron A, et al. Crossing the edge: Determinants of movements. In: Andersson AA, de Garine-Wichatitsky M, DHM C, Dzingirai V, Giller KE, editors. Transfrontier Conservation Areas: People living on the edge. London: Routledge; 2013. 123-136 pp

[10] Pretty J. Social capital and connectedness: Issues and implications for agriculture, rural development and natural resource management in ACP countries. Review Paper for CTA. CTA Working Document Number 8032. 2003

[11] Chema S, Gilbert E, Roseboom J. A critical review of key issues and recent experiences in reforming agricultural research in Africa. In: Draft Working Paper. The Hague: ISNAR; 2002

[12] Mudzengi CP, Taderera LM, Tigere A, Kapembeza CS, Moyana S, Zimondi M, et al. Adoption of urea treatment of maize stover technology for dry season supplementation of cattle in Wedza, Zimbabwe. Livestock Research for Rural Development. 2014;**26**:160. Available from: http://www.lrrd.org/lrrd26/9/mudz26160.htm

[13] Chenje M, Sola L, Paleczny D. The State of Zimbabwe's Environment. Harare: Government of the Republic of Zimbabwe Ministry of Mines, Environment and Tourism; 1998

[14] Nyamudeza P, Hussein J, Matibiri B. The sustainable management of vertisols. In: Syers JK, Penning de Vries FWT, Nyamudeza P, editors. Vertisols Management in Zimbabwe. Wallingford: Cabi International; 2001

[15] Murwira A. Scale matters: A new approach to quantify spatial heterogeneity for predicting the distribution of wildlife [doctoral thesis]. Wageningen University; 2003

[16] Bryant DM, Ducey MJ, Innes JC, Lee TD, Eckert RT, Zarin DJ. Forest community analysis and the point-centered quarter method. Plant Ecology. 2005;**175**:193-203

[17] Carruthers V. The Wildlife of Southern Africa: A Field Guide to the Animals and Plants of the Region. Western Cape: Southern Book Publishers; 1997

[18] Plower DCH, Drummond RB. Wild Flowers of Zimbabwe (Revised Edition): A Guide to some of the Common Wild Flowers of Zimbabwe. Harare: Zimbabwe, Longman Publishers; 1990

[19] Drummond RB. A list of trees, shrubs and woody climbers indigenous or naturalised in Rhodesia. Kirkia. 1975;**10**(1):267

[20] Walker BH. An approach to the monitoring of changes in the composition and utilisation of woodland and Savanna vegetation. Southern African Journal of Wildlife Resources. 1976;**6**(1):1-32

[21] Rattray JM. The grass and grass associations of Southern Rhodesia. Rhodesia Agricultural Journal. 1957;**54**:197-234

[22] Ngongoni NT, Mapiye C, Mwale M, Mupeta B. Factors affecting milk production in the smallholder dairy sector in Zimbabwe. Livestock Research for Rural Development. 2006;**18**. Available from: http://www.cipav.org.co/lrrd/lrrd18/5/ngon18072.htm

[23] Dzavo T, Zindove TJ, Dhliwayo M, Chimonyo M. Effects of drought on cattle production in sub-tropical environments. Tropical Animal Health and Production. 2019;**51**:669. DOI: 10.1007/s11250-018-1741-1

[24] Makwara C. Sustainable and profitable farming through conservation agriculture in Zimbabwe: Prospects, opportunities and constraints. Journal of Sustainable Development in Africa. 2010;**12**:181-190

[25] Makhado RA, Mapaure I, Potgieter MJ, Luus-Powell WJ, Saidi AT. Factors influencing the adaptation and distribution of Colophospermum mopane in Southern Africa's mopane savannas—A review. Bothalia. 2014;**44**:9

[26] Mlambo V, Smith T, Owen E, Mould FL, Sikosana JLN, Mueller Harvey I. Tanniniferous *Dichrostachys cinerea* fruits do not require detoxification for goat nutrition: In sacco and in vivo evaluations. Livestock Production Science. 2004;**90**:135-144

[27] Bamigboye F, Babayemi O, Adekoya A. Feed resources and seasonal nutrient composition of predominant forages for small ruminant production in Iwo local government area of Osun State, Nigeria. Journal of Biology, Agriculture and Healthcare. 2013;**3**:15-24

[28] Abusuwar AO, Ahmed EO. Seasonal variability in nutritive value of ruminant diets under open grazing system in the semi-arid rangeland of Sudan (South Darfur State). Agriculture and Biology Journal of North America. 2010;**1**:243-249

[29] Murungweni C, Van Wijk MT, Andersson JA, Smaling EMA, Giller KE. Application of fuzzy

cognitive mapping in livelihood vulnerability analysis. Ecology and Society. 2011;**16**:8

[30] Madibela OR, Seitshiro O, Mochankana ME. Deactivation effects of polyethylene glycol (PEG) on in vitro dry matter digestibility of *Colophospermum mopane* [Mophane] and Acacia browse trees in Botswana. Pakistan Journal of Nutrition. 2006;**5**:343-347

[31] Codron D, Lee-Thorp JA, Sponheimer M, Codron J. Nutritional content of savanna plant foods: Implications for browser/grazer models of ungulate diversification. European Journal of Wildlife Research. 2007;**53**:100-111

[32] Lucena RFP, Araújo EL, Albuquerque UP. Does the local availability of woody Caatinga plants (Northeastern Brazil) explain their use value? Economic Botany. 2007;**61**:347-361

[33] Banso A, Adeyemo O. Evaluation of antibacterial properties of tannins isolated from Dichrostachys cinerea. African Journal of Biotechnology. 2007;**6**:1785-1787

[34] Gondo T, Frost P, Kozanayi W, Stack J, Mushongahande M. Linking knowledge and practice: Assessing options for sustainable use of mopane worms *[Imbasiabelina]* in Southern Zimbabwe. Journal of Sustainable Development in Africa. 2010;**1**:281-305

[35] Rusinga O, Maposa R. Traditional religion and natural resources: A reflection on the significance of indigenous knowledge systems on the utilisation of natural resources among the Ndau people in South Eastern Zimbabwe. Journal of Ecology and the Natural Environment. 2010;**2**:201-206

[36] Mudzengi CP, Murwira A, Zengeya FM, Murungweni C. Screening key browse species in a semi-arid rangeland. Cogent Food and Agriculture. 2017;**3**:1285854

[37] Na R, Dong H, Tao X, Ma R, Xi J. Effects of diet composition on in vitro digestibility and methane emissions of cows. Journal of Agro-Environment Science. 2010;**29**(8):1576-1581

[38] Ngongoni NT, Mapiye C, Mupeta B, Mwale M, Chimonyo M. Potential of farm-produced crop residues as protein sources for small-medium yielding dairy cows. African Journal of Agricultural Research. 2007;**2**:309-317

[39] Mapiye C, Mupangwa JF, Mugabe PH, Chikumba N, Poshiwa X, Foti R. A review of forage legume research for rangeland improvement in Zimbabwe. Tropical Grasslands. 2006;**40**:145-149

[40] Giller KE. Nitrogen fixation in the Tropical Systems. 2nd ed. Wallingford, UK: CABI; 2001. 150-300 pp

[41] Longland AC. Pastures and pasture management. In: Geor RL, Harris PA, Coenen M, editors. Equine Applied and Clinical Nutrition. Edinburgh, New York: Saunders Elsevier; 2013

[42] Quaas MF, Baumgärtner S. Optimal grazing management rules in semi-arid rangelands with uncertain rainfall. Working Paper Series in Economics, No. 193. University of Lüneburg; 2011

[43] Muller B, Frank K, Wissel C. Relevance of rest periods in non-equilibrium rangeland systems: A modelling analysis. Agricultural Systems. 2007;**92**:295-317

[44] Quaas MF, Baumgartner S, Becker C, Frank K, Muller B. Uncertainty and sustainability in the management of rangelands. Ecological Economics. 2007;**62**:213-234

[45] Wairore JN, Mureithi SM, Wasonga OV, Nyberg G. Benefits derived from rehabilitating a degraded semi-arid rangeland in private enclosures in West Pokot County, Kenya. Land Degradation and Development. 2016;**27**(3):532-541. DOI: 10.1002/ldr.2420

[46] Stephen M, Mureithi SM, Verdoodt A, Njoka JT, Gachene CKK, van Ranst E. Benefits derived from rehabilitating a degraded semi-arid rangeland in communal enclosures, Kenya. Land Degradation and Development. 2016;**27**(8):1853-1862. DOI: 10.1002/ldr.2341

Section 3

Livestock Farming and Climate Change

Chapter 7

Climate Change Mitigation in Livestock Production: Nonconventional Feedstuffs and Alternative Additives

Pámanes-Carrasco Gerardo, Herrera-Torres Esperanza, Murillo-Ortiz Manuel and Reyes-Jáquez Damián

Abstract

Livestock production has widely contributed to increase global production of greenhouse gases (GHG), mostly through digestive fermentation in ruminants. Moreover, emissions derived from livestock are estimated over 14% of the total anthropogenic GHG emissions to atmosphere. In addition, methane emitted from ruminal enteric fermentation is responsible for 25% of the total global methane emissions, which turns livestock activity into a main promoter of the climate change effect. However, these emissions may be diminished by modifying livestock diets through alterations in forage-concentrate ratios, the supplementation of feed additives, and the inclusion of alternative feedstuffs not commonly used as forage and protein sources in farm animal feeding. Additionally, the use of nonconventional feedstuffs is highly recommended since their production does not compete with human feeding and may provide metabolites used as methanogenesis suppressors. Likewise, agricultural by-products should be considered as potential feedstuffs for animal production by increasing the livestock efficiency and reducing the energy losses due to methane synthesis.

Keywords: methanogenesis inhibitors, nonconventional forages, feedstuff additives, secondary metabolites, methanogens

1. Introduction

The world's population have substantially increased in the last decades, and it is expected to keep increasing for the next 30 years until the population reaches 9.8 billion in 2050 [1]. Consequently, there is a growing demand for food and natural resources for human surviving. Livestock represents a main source of protein and energy for human consumption, as well as an important basis of financial revenues for families at rural areas. However, this economic activity is positively correlated to the climate change (CC) effect [2]. In the last centuries, CO_2 and NO_2 emissions have increased 31 and 16%, respectively; whereas, methane has increased twofold. Approximately, 40% of the methane emitted to atmosphere is originated from natural sources [2]; the remaining 60% is originated from anthropogenic sources (livestock, rice crops, fossil fuel exploitation, and dump).

In this way, emissions derived from livestock are estimated over 14% of the total anthropogenic greenhouse gases (GHG) emitted to atmosphere, which account approximately 50 gigatons of CO_2 equivalents per year (GTons-CO_2 equiv./yr) [3]. In addition, livestock is a major non-CO_2 GHG producer (CH_4 and NO_2); these gases possess a higher trapping heat index compared to at least 25 times for CO_2 [3].

Climate change effect has risen the average planet temperature approximately 1°C. In fact, polar caps are melting rapidly, which have increased the sea levels as a consequence [3]. If these trends keep on going, the CC effect will reach a non-return point, causing irreparably damages to the planet [4]. In addition, the UN encouraged developing countries (mainly Latin American countries) to strengthen their efforts to avoid an increase over 1.5°C in the temperature of the planet. Nevertheless, since CO_2 emissions increased substantially in the latest years, a 3°C rise of the temperature is expected by the end of the century [5].

Due to the latter, worldwide researchers and governments attempt to mitigate livestock gases production by changing livestock diets and offering alternative feedstuffs as an important strategy to mitigate GHG emissions and CC effect.

2. Ruminal enteric fermentation and methanogenesis

Methanogenesis was once considered a singular type of fermentation. However, in some respects, a very unique biochemistry is involved. The process is carried out by strictly anaerobic bacteria, all of which belong to the phylum *Euryarchaeota* in five orders that include mesophiles and thermophiles: *Methanobacteriales*, *Methanococcales*, *Methanomicrobiales*, *Methanopyrales*, and *Methanosarcinales*. Methanogens can be found in freshwater and marine environments, cold

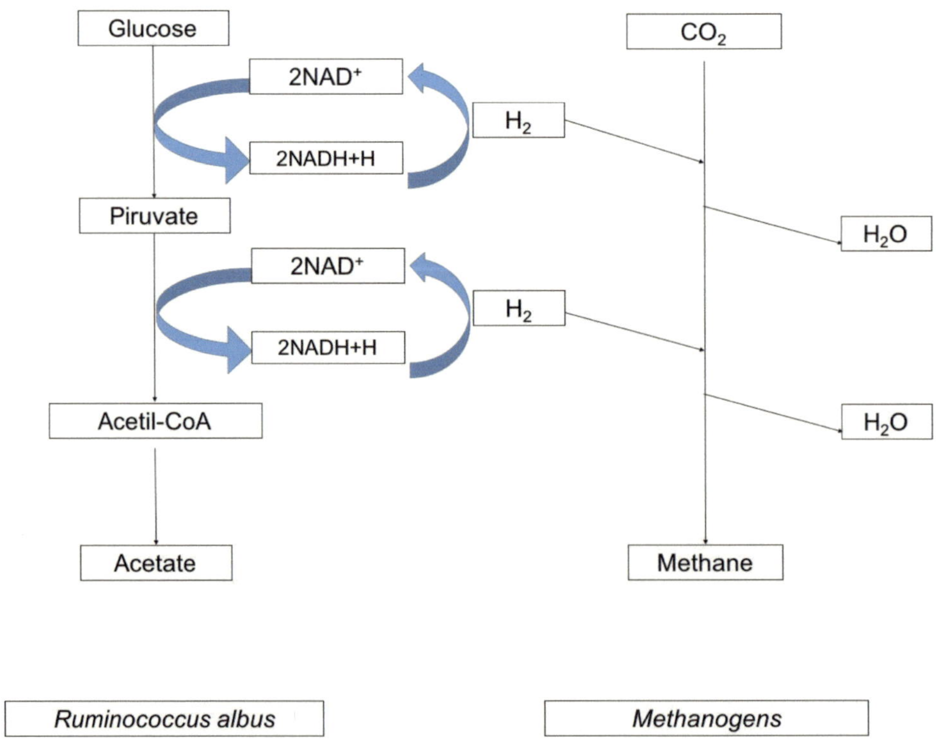

Figure 1.
Synergic relationship between Ruminococcus albus and methanogens (adapted from [10]).

sediments, and hydrothermal vents as free cells living in symbiosis within animals which produce methane as well as in symbiosis with anaerobic methane oxidation-promoting bacteria [6].

Ruminal degradation of fiber and starch generates hexoses which later are fermented through the glycolysis pathway. Pyruvate, as a final product of the glycolysis, is converted into volatile fatty acids (VFA), mainly acetic, propionic, and butyric acid, through different metabolic pathways. These VFA are rapidly absorbed by the animal and are used as energy source, while other products such as H_2 and CO_2 are generated. However, the hydrogen produced in the glycolysis inhibits $NADH+H^+$ ferredoxin oxidoreductase enzyme, which impedes NAD regeneration when a low H_2 pressure is present [7]. Therefore, methane production is essential for obtaining a high-performing rumen ecosystem, because H_2 accumulation is avoided, which could then inhibit dehydrogenase activity in later re-oxidation cofactors. An efficient H_2 capture in the rumen contributes to increase the rate of fermentation by the lack of its inhibitory effect on the microbial degradation of vegetative material [8, 9]. Hence, thermodynamically methane synthesis is favored. **Figure 1** represents the synergic relationship between *Ruminococcus albus* and methanogens, as an example of the expressed earlier [10].

3. Fermentation modifiers

The rumen is an anaerobic bioreactor which contains a great diversity of microorganisms, such as bacteria, fungi, protozoa, and archaea. From all of these, just a few have been cultivable and virtually identified. However, the newer molecular biology techniques have widely contributed to the identification of ruminal microorganisms, as well as the activity from each consortium in the ruminal fermentation. Feedstuffs' degradation in the rumen is effectuated by microorganisms with different goals and at different proportions. In addition, the enzymatic and degradative activity of every consortium may be affected by several factors, such as diet, season, inherent characteristics of the ruminant's breed, geographic zone, feeding strategies, physiological conditions, intake, etc. [11]. Hence, modification in the ruminal fermentation can be achieved by alterations on the previously mentioned variables, showing positive changes in efficiency and productivity of the animal. Therefore, diverse targets have been defined through modification in the ruminal fermentation: (a) to decrease the ruminal methane synthesis through the increase of propionate production; (b) to improve fibers' ruminal digestion; (c) to increase undegradable rumen protein in order to increase the bypass protein to lower tract which later will be absorbed by the animal through the intestine walls; and (d) to reduce rapidly degradable carbohydrates in rumen [12]. According to the latter, diverse options have been studied to cover two or more targets.

3.1 Nonfibrous carbohydrates

Carbohydrate fermentation is the main source of energy for the ruminant. Quantity and quality of rapidly degradable carbohydrates, usually known as nonfibrous carbohydrates (NFCs), depends on the feedstuff. Thus, NFCs contained in corn (*Zea mays*) are mostly starch, whereas, in molasses, NFCs are mainly composed by mono- and disaccharides. Depending on the NFC type and the supplied feedstuff, certain pathways for synthesis may be favored. For example, whether increases in the structural carbohydrates are observed, the propionate synthesis pathway is enhanced. This pathway is beneficial to the animal since it reduces methane synthesis [13]. Otherwise, an increase in mono- and disaccharides decreases microbial protein synthesis through reductions in the abundance of ammonia-utilizing cellulolytic bacteria [14]. Moreover, high NFC concentrations tend to increase VFA production which could cause ruminal acidosis.

3.2 Fibrous carbohydrates

It has been demonstrated that increases in dry matter intake reduce methane production [15]. Moreover, increases in digestibility is expected in fibrous material whether it is fine ground, as well as augmentations in the passage rate through increases in the turnover rate. Therefore, if turnover rate is increased, the passage rate would also increase. Hence, through augmentations in the passage rate, microorganisms that possess a lower growth rate, such as protozoa and archaea, will defaunate, thus decreasing methane production [16]. Otherwise, digestibility and methane production could be increased by increasing the retention time [17]. Additionally, by increasing the intake above the minimum for maintenance, the animal methane production will arise proportionally. This phenomenon will provoke a reduction in methane production per production unity [18]. Therefore, an animal fed under a pasture basis will produce less methane as part of the GHG produced compared to an animal fed with a high-concentrate or high-fiber proportion diet.

3.3 Bypass protein

The protein contained in ruminants' feedstuffs could be divided into two groups: degradable rumen protein (DRP) and undegradable rumen protein (URP). The first is degraded in rumen, and it is used as a nitrogen source in the microbial protein synthesis; the second escapes from ruminal degradation and is transported to the lower tract where it is susceptible of being absorbed by the animal in the form of amino acids [19]. In spite of several reasons to name it bypass protein, one of the main characteristics is its low retention time in rumen or, the inverse action, the high passage rate. In the case of high passage rates, microorganisms which possess a low growth rate will tend to defaunate; this is the case of the methanogens. Thus, methanogenesis will be affected and methane production will be reduced. Nowadays, some secondary metabolites are identified as protein protectors, by forming complexes with proteins and avoiding their degradation in rumen. The latter allows proteins to go through the low tract and to be absorbed after liberating complexes due to the acidic pH in the intestine [20].

4. Feed additives

Some strategies are focused on providing feed additives to modify the presence or absence of methanogens, protozoa, or the direct or indirect inhibition of ruminal methanogenesis [21]. By supplementing feed additives, good results are observed in methane production and productive performance. These strategies imply the use of high nutritive quality forages, organic acids, ionophores, probiotics, vegetable extractives, and secondary metabolites from different plants [22]. However, the most used are presented and briefly discussed:

- *Ionophores*: Ionophores are additives which possess a proved antimicrobial effect on some ruminal and cultivable strains, especially gram-positive bacteria [23]. Ionophore compounds like monensin and lasalocid have demonstrated to modify rumen fermentation and decrease methane emissions. The latter can be elucidated due to the fact that ionophores, as mentioned earlier, present affinity to hydrogen- and formate-producing, butyrate-producing, lactate-producing, and ammonia-producing bacteria, all of them gram-positive. However, succinate- and propionate-producing bacteria are resistant to ionophores [24]. Hence, it is assumed that reductions in the methanogenesis pathway are due to the hydrogen capture by propionate-producing bacteria, limiting methanogenesis through the restriction of hydrogen availability in the CO_2 reduction pathway. Unfortunately, prolonged use of monensin in steers has shown a loss of methanogenesis inhibition action and a resistance of bacteria to these antibiotics [25].

- *Homoacetogens*: Homoacetogens are a group of acetate-producing bacteria which can convert carbon dioxide into acetate using hydrogen [26]. The acetogenesis is a competitive pathway against methanogenesis for hydrogen use. Additionally, the production of ruminal acetate can be used as an energy source for the animal [27]. However, the thermodynamics of the reactions are more favorable to methanogenesis, and the use of ruminal homoacetogens as additives did not suppress methanogenesis in all the studies [28, 29].

- *Essential oils*: The effect of the addition of some essential oils into methanogenesis is through the capture of hydrogens in the biohydrogenation process of unsaturated fatty acids in the rumen [30]. Likewise, some medium-chain fatty acids contained in vegetable oils have demonstrated suppression of methanogenesis through the reduction of methanogens and ciliate [31]. In addition, some authors stated that the methanogenesis suppression with coconut oil was due to a change in methanogens population [32].

- *Yeast cultures*: The most used yeast culture in livestock research is *Saccharomyces cerevisiae*, and it has been used as a fermentation modifier [23]. Additionally, yeast cultures have been used as rumen fermentation modifiers and promoters of microbial growth [33]. In fact, rumen fibrolytic bacteria have a clear preference for a nitrogen source for ammonia production, and this is enhanced by yeast cultures for microbial protein synthesis [34]. Moreover, recent reports have suggested the stabilization of pH through a decrease in lactate production when using in vitro yeast cultures [35]. Thus, the antimethanogenic action is suggested through the improvement of fiber digestion and increasing ammonia-utilizing bacteria [36].

- *Others*: Vaccination and the use of bacteriophages are a different alternative for methane mitigation. Hence, vaccines against methanogens like *Methanobrevibacter* spp. have been applied to sheep presenting methane reductions of 7.7% [37]. Likewise, the use of phagaes against rumen archaea has been suggested by other authors as a strategy for methane abatement [22].

5. Conventional and nonconventional forage sources

As expressed before in this chapter, the increasing global population demands for a higher feed production, converting animal feeding production into a natural competitor for human feeding production in the search for arable lands. Consequently, diverse researches have focused into trying different forage sources which were not conventional as animal feeding before but now could be considered as alternative forage sources [38, 39]. Nonconventional forages include a wide variety of perennial plants and agriculture and commercial by-products which do not compete with human feeding. Therefore, diverse advantages can be observed when utilizing alternative forages such as (a) a considerable reduction in the feeding source costs; (b) exploitation of nutrients contained in agriculture by-products which otherwise would not be used (these by-products could be high in rapidly digestible carbohydrates or in fiber, both suitable for ruminants); and (c) an increase in the by-product cost which eventually will create economic benefits for producers and the productivity chain supply. Moreover, some agricultural by-products do not need any processing to be offered as animal feeding, hence the desirability of these by-products. Some of the ruminants feeding produced under this basis are:

a. Crop by-products such as garlic leaves, onion leaves, cocoa husks, coconut meal, cracked rice, sugarcane bagasse, molasses, tapioca discards, oat straws, and some aquatic crops like water hyacinth and azolla [38–41]

b. Perennial crops, seeds, and leaves of shrubs and trees like *Leucaena*, guamuchil, mesquite, mango, ebony, etc. [42–44]

Some farmers are still not aware of the nutritional value contained in by-products or in the form to be included into the productivity chain of animal feeding in an efficient way. In this regard, Asia and Africa are heavily focused on attempting to reach this goal. Due to the nature of some agricultural by-products, these tend to decompose in a short time. Hence, some techniques should be used to preserve and increase their shelf life. Therefore, some of the preservation techniques commonly used are listed as follows:

I. *Silages*: Many of the agricultural by-products are obtained in huge quantities due to the nature of the crops. However, the high humidity contents contribute to a short lifetime due to the rapidly appearance of fungi and, eventually, a decomposition. Therefore, the silage elaboration is a recommended preservation method due its large periods of storage, and it can always be offered fresh and with certain aroma provided from the fatty acids synthetized in the lactic fermentation which will add palatability for ruminants [45].

II. *Chemical treatments*: Some agricultural residues obtained from cereal crops are treated with chemicals to increase their digestibility. Thus, by-products with high lignocellulosic complexes could be treated with ammonia in anaerobic conditions to enhance lignin and fiber hydrolysis, which will improve their digestibility [46]. However, this process requires special plastic sheets that increases costs and could become an unaffordable process for small producers. In this way, previous researches have reported diverse alternatives using mud and eliminating the use of plastic sheets [47].

III. *Multi-nutritional blocks*: Another conservation technique which involves the utilization of high humidity agricultural by-products is the elaboration of multi-nutritional blocks [48]. This technology is very flexible and allows the producer to use ingredients considered as indispensable in animal feeding. Additionally, important nutrients could be available for longer periods of time since the useful life of these blocks is very extensive. Although in dry seasons, drought decreases considerably the nutritional quality of forages. These blocks are generally offered as supplementation in livestock feeding in rangelands as part of an extensive feeding system, and they are commonly elaborated to supplement vitamins and minerals, such as zinc and copper [49]. However, recent investigations are using more ingredients to improve their nutritional value and turn this into a more versatile practice.

IV. *Supplementation*: Supplementation is extensively used, especially if there is a deficient feed due to poor nutritional quality of some ingredients. By using this technique, some essential nutrients will be delivered to the livestock which otherwise could not be obtained by the animal itself. Nevertheless, the acquisition of ingredients for supplementation is unaffordable for some small producers. On the other hand, there are certain agricultural by-products which could be offered to the livestock and contain certain important nutrients at very low or even null cost. In this way, shrub and tree seeds could be a very good option. *Leucaena*, guamuchil, and ebony seeds are rich in protein and unsaturated lipids; these are being used without any affections in productive performance in small ruminants at very low cost. In this way, shrubs and trees seeds could be a very good option. Leucaena, ***Vicia faba***, ebony and other seeds are rich in protein and unsaturated lipids; these are being used without any affections in productive performance in small ruminants at a very low cost [50–53].

6. Plant metabolites

In the last years, ruminants have been target of several feeding strategies aiming to reduce ruminal methane production and emissions; most of them have been stated earlier in this chapter. However, the use of secondary metabolites arises as a viable and newer alternative in this concern. There is evidence which proves certain secondary metabolites, such as condensed tannins, saponins, and alkaloids, reduce methane production in in vivo and in vitro assays [54]. Generally, the mechanisms of action of these compounds point out to certain metabolic pathways:

I. **Tannins**: Tannins are water-soluble polyphenol polymers with a high and diverse molecular weight. They can form complexes with proteins, mainly, and metal ions, amino acids, and polysaccharides in a lesser extent. These metabolites are normally synthetized in shrubs, trees, legumes, fruits, cereals, and grains [55]. Tannins are divided into two groups: condensed tannins (CTs) and hydrolysable tannins (HTs).

Hydrolysable tannins: These are complex molecules attached to a polyol group as a central core which are partially or fully esterified with a phenolic group (e.g., gallic acid). The remaining phenolic groups could be later esterified or oxidized to produce more complexes with HTs [55].

Condensed tannins: These compounds are also known as proanthocyanidins and are mainly polymers of the flavan-3-ol units which are bind by interflavonoids C3-C8 and C4-C6 linkages, such as catechin and epicatechin. The methanogenic activity conferred to tannins is mostly due to the condensed tannins; CTs attach to proteins and avoid their degradation in rumen. Additionally, CT decrease methanogenesis through a reduction in fiber digestion [56]. Some studies affirm that CT enhances acetate formation via acetogenesis; this metabolic pathway uses hydrogen for acetate synthesis and reduces methanogenesis [57].

II. **Saponins**. According to their chemical structure, they are divided into two groups: steroids and triterpenoids. Steroids are predominantly in plants and are composed of 27 carbon atoms in the central skeleton of its molecule (e.g., spirostanol and furostanol). Otherwise, triterpenoids are composed mainly of aglycones with 30 carbon atoms in its molecule (e.g., oleanane) [58]. These are the most common types of saponins, especially in legumes [59]. Methanogenic action of saponins occurs by protozoa defaunation which is associated to methanogens. Moreover, saponins enhance production of propionate, a natural competitor of methane in hydrogen capture [58]. Nevertheless, some studies affirm that methane inhibition action by saponins is dose and time dependent and not conclusive [59].

III. **Flavonoids**. Flavonoids are phenolic compounds (like tannins); however, these contain only 15 carbon atoms linked to 2 aromatic rings connected through a 3-carbon bridge [60]. These metabolites are particularly studied for human purposes, and their biological benefits to health correlated to their consumption [61]. Almost all flavonoids are conjugated to glycosides and are common to find hydroxyl groups in carbons with four, five, and seven positions [60]. In addition, flavonoids stimulate microbial metabolism and reduce methane production through enhancing acetogenesis pathway and increasing hydrogen capture in propionate anabolism, in a similar way as described earlier with saponins [28, 59].

7. Other feedstuffs

Since the 1970s, ruminal microbes and their effect on ingested nutrients have been subject of intensive research [27]. Ruminal microorganisms are crucial for the digestive performance of animals. Addition of feedstuffs in diets of ruminants has led to investigate their effects on the absorption and utilization of nutrients as well as the ruminal environment and conditions. Genetically modified *Escherichia coli* showed a ruminal methanogenesis mitigation effect in sheep [62]. Other researches [63] reported that *Lactococcus lactis* produces nisin, which has demonstrated antimicrobial activity against Gram-positive bacteria, resulting in a mitigation effect on ruminal methane emission.

β1–β4 galacto-oligosaccharides (GOS), along with glucose, fructose, and starch, present in the rumen are used by *Bifidobacterium* and *Lactobacillus* as substrates to produce lactate and acetate. Lactate is one of the main transitional compounds during propionate production, which competes against methanogens for available hydrogen. As a result, methane production can be decreased by GOS consumption [64].

7.1 Probiotics

Probiotics are commonly defined as "live micro-organisms which, when administered in adequate amounts, confer a health benefit on the host." Other authors indicate that a probiotic food carries 106–107 CFU/g viable probiotic cells, until the shelf life of the product is reached [65]. Probiotic foods contain sensitive ingredients, such as probiotic cells that require protection against oxidative stress, high acidity, freezing, shear stress, and other undesirable factors. Although microencapsulation has been primarily used to protect bioactive ingredients due to its advantages [66], co-extrusion technology has become an emerging alternative to encapsulate probiotic bacteria.

8. Extrudates and extrusion process

8.1 The use of the extrusion process in the supplementation of probiotics

Extrusion processing using oil and alginate solutions to create emulsions as core medium [67] has found a favorable survival of probiotic *L. acidophilus* at 4°C for 50 days. Over the years and because of technological advances, extrusion has become an almost unlimited cooking processing alternative due its inherent versatility. Multiple studies had focused on designing and evaluating the incorporation of biomass, distillery by-products, fruit pomaces, agro-industrial by-products, and dairy residues [68]. One of the main advantages of the thermal and pressure conditions during extrusion is the inactivation of antinutritional factors, elimination of pathogens, improved digestibility, reduced level of toxins, as well as the bitterness of some oil plants (flax, cotton, peanut, and sunflower) while achieving the desired organoleptic characteristics by properly adjusting residence times, specific energy absorbed, and pressure effects on the raw materials [67]. Other authors extruded rye whole meal to decrease microbial contamination and used it as cultivation medium for the evaluation of supplementation of dairy cow ration with *P. pentosaceus* BaltBio02 (9.6 log10 CFU g^{-1} $head^{-1}$ day^{-1}) [69]. Obtained results showed an increase ($P < 0.05$) of milk yield but did not affect milk composition or ruminal fermentation parameters. *Lactobacillus sakei* KTU 05-6 (9.6 log10 CFU g^{-1} $head^{-1}$ day^{-1}) was also analyzed but showed no significant impact on yield or ruminal parameters.

On the other hand, a different study evaluated the effect of different doses of probiotic containing 1.6 × 109 CFU/g of *Bacillus licheniformis* and 1.6 × 109 CFU/g of *Bacillus subtilis* on in vitro digestibility of concentrates and forages [70]. These authors concluded that 3 g $head^{-1}$ d^{-1} of probiotic increased by 10.9% starch digestibility after 12 h of incubation, indicating a promotion of NDF digestibility in roughages and starch in concentrates, although no significant changes were obtained of acetate, propionate, and butyrate molar ratios, possibly due to negligible changes on H^+ concentrations that affect the environmental pH of ruminal microorganisms [71]. An enhanced VFA production results in a pH reduction and growth inhibition of fermenting fibrous carbohydrate bacteria, which compromise NDF digestibility.

9. Current strategies

9.1 Methane reduction through improvement of the forage quality

There is a lot of information about supplementation of secondary metabolites, certain additives, and increasing concentrate fraction in the diet of livestock to abate methane emissions. However, some producers in developing countries are not able to afford these alternatives. Otherwise, methane production in ruminants in developing countries is directly correlated to a poor quality in feedstuffs offered to livestock, by decreasing the efficiency and productivity for productive unit [72]. In this way, the strategies that producers and researchers in developing countries use imply the production of improved forage sources which is cheaper than the acquisition of some supplements. Additionally, the use of these forage sources may increase the fertility in the soil which is desirable for nitrogen fixation. Consequently, by improving the quality and quantity of forage, the productivity will increase, and methane production will be reduced by productive unit.

9.2 Vaccination and chemical compounds

On the other hand, other researchers have focused their efforts on evaluating the inclusion of protected lipids and nitrate compounds [73, 74]. In addition, the use of some nitrate compounds showed no effect on organoleptic and nutritional properties in edible products for ruminants [75]. However, both strategies could be discarded by increases on fiber digestibility and a reduction of dry matter intake. Otherwise, the acquisition and use of these compounds in livestock will substantially increase production costs and market price. In the past decades, chemical compounds were used as inhibitors in methane synthesis through vaccination or the analogue supplementation. Nevertheless, methanogen defaunation is not a viable long-term alternative since microorganisms are easily adaptable to different environments. Additionally, the use of other additives, like ionophores, is forbidden in the USA. In this way, the use of plant extractives and especially metabolites arises as a sustainable alternative; however, there are not conclusive results which lead to a punctual design of dietary strategies. The latter is exposed since some of these metabolites may be present in edible products of ruminants affecting their organoleptic properties [76]. In addition, further studies are required to demonstrate the effectivity of extractable compounds of plants which are well perceived by the population as an alternative for chemical compound supplementation.

10. Conclusions

Methane and GHG mitigation in livestock is possible through different strategies, most of them as dietary alterations. However, it is necessary to carry out conclusive in vivo studies evaluating the use of metabolites and extractable plants' compounds, as well as the use of alternative forage sources which may provide directly these metabolites affecting the presence of ruminal methanogens and protozoa. Moreover, each region or geographic zone has different forage sources even perennial that can be produced locally. The incorporation of these into livestock feeding arises as a viable and sustainable alternative for mitigating GHG emissions, especially methane.

Acknowledgements

The authors would like to acknowledge the National Council of Science and Technology (CONACYT) for indirect support of some researchers. Likewise, the authors would like to acknowledge the Forestry and Wood Industry Institute and Veterinary Medicine and Husbandry Faculty of the UJED, as well as the Durango Institute of Technology and Technological Institute of the Valle del Guadiana for the facilities.

Author details

Pámanes-Carrasco Gerardo[1]*, Herrera-Torres Esperanza[2], Murillo-Ortiz Manuel[3] and Reyes-Jáquez Damián[4]

1 CONACYT, Durango State Juarez University, Durango, Dgo, Mexico

2 TecNM, Valle del Guadiana Institute of Technology, Durango, Dgo, Mexico

3 Durango State Juarez University, Durango, Dgo, Mexico

4 TecNM, Durango Institute of Technology, Durango, Dgo, Mexico

*Address all correspondence to: gerardo.pamanes@gmail.com

IntechOpen

© 2019 The Author(s). Licensee IntechOpen. This chapter is distributed under the terms of the Creative Commons Attribution License (http://creativecommons.org/licenses/by/3.0), which permits unrestricted use, distribution, and reproduction in any medium, provided the original work is properly cited.

References

[1] Alexandrato N, Bruinsma J. World agriculture towards 2030/2050: The 2012 revision. In: ESA Working Paper. Vol. 12(03). Rome: FAO; 2012

[2] WMO. WMO Greenhouse Gas Bulletin No. 14: The State of Greenhouse Gases in the Atmosphere Based on Global Observations through 2017. Geneva, Switzerland: World Meteorological Organization; 2018

[3] Intergovernmental Panel on Climate Change (IPCC). Summary for policymakers. In: Climate Change 2014, Mitigation of Climate Change. Contribution of Working Group III to the Fifth Assessment Report of the Intergovernmental Panel on Climate Change. 2014

[4] UN. World 'Nearing Critical Point of No Return' on Climate Change, Delegate Warns, as Second Committee Debates Sustainable Development. In: Second Committee Seventy-Third Sesión, United Nations. New York: USA. October 2018

[5] Butler JH, Montzka SA. The NOAA Annual Greenhouse Gas Index (AGGI). Published online Spring 2015. Retrieved October 5, 2015, from: http://www.esrl.noaa.gov/gmd/aggi/aggi.html

[6] Fenchel T, King GM, Blackburn TH. Microbial biogeochemistry and extreme environments. In: King G, editor. Bacterial Biogeochemistry. Elsevier Ltd; 2012. pp. 143-161. DOI: 10.1016/B978-0-12-415836-8.00008-6

[7] Hegarty RS, Gerdes R. Hydrogen production and transfer in the rumen. Recent Advances in Animal Nutrition. 1999;**12**:37-44

[8] Ribeiro LG, Machado FS, Campos MM, Guimaraes R, Tomich TR, Reis LG, et al. Enteric methane mitigation strategies in ruminants: A review. Revista Colombiana de Ciencias Pecuarias. 2015;**28**:124-143. DOI: 10.17533/udea.rccp.v28n2a02

[9] McAllister TA, Newbold CJ. Redirecting rumen methane to reduce methanogenesis. Australian Journal of Experimental Agriculture. 2008;**48**: 7-13. DOI: 10.1071/EA07218

[10] Moss AR, Jouany JP, Newbold J. Methane production by ruminants: Its contribution to global warming. Annals of Zootechnia. 2000;**49**(3):231-254

[11] Hook SE, Wright A, Mcbride BW. Methanogens: Methane producers of the rumen and mitigation strategies. Archaea. 2010:945785. DOI: 10.1155/2010/945785

[12] Morgavi DP, Forano E, Martin C, Newbold CJ. Microbial ecosystem and methanogenesis in ruminants. Animal. 2010;**4**(7):1024-1036. DOI: 10.1017/S1751731110000546

[13] Ferraro S, Mendoza G, Miranda A, Gutierrez C. In vitro gas production and ruminal fermentation of glycerol, propylene glycol and molasses. Animal Feed Science and Technology. 2009:112-154. DOI: 10.1016/j.anifeedsci.2009.07.009

[14] Bach A, Calsamiglia S, Stern MD. Nitrogen metabolism in the rumen. Journal of Dairy Science. 2005;**88**(E9-E21):73133-73137. DOI: 10.3168/jds.S0022-0302

[15] Molano G, Clark H, Knight TW, Cavanagh A. Methane emissions from growing beef cattle grazing hill country pasture. Proceedings of the New Zealand Society of Animal Production. 2006;**66**:172-175

[16] Hegarty RS. Reducing rumen methane emissions through elimination

of rumen protozoa. Australian Journal of Agricultural Research. 1999;**50**: 1321-1327. DOI: 10.1071/AR99008

[17] Pinares-Patino CS, Baumont R, Martin C. Methane emissions by Charolais cows grazing a monospecific pasture of timothy at four stages of maturity. Canadian Journal of Animal Science. 2003;**83**(4):769-777. DOI: 10.4141/A03-034

[18] Henry B, Charmley E, Eckard R, Gaughan J, Hegarty R. Livestock production in a changing climate: Adaptation and mitigation research in Australia. Crop and Pasture Science. 2012;**63**:191-202. DOI: 10.1071/CP11169

[19] Tandon M, Siddique RA. Role of bypass proteins in ruminant production. Dairy Planner. 2016;**4**(10):11-14. DOI: 10.13140/RG.2.2.16615.04003

[20] Newbold CJ, de la Fuente G, Belanche A, Ramos-Morales E, McEwan NR. The role of ciliate protozoa in the rumen. Frontiers in Microbiology. 2015;**6**:1313. DOI: 10.3389/fmicb.2015.01313

[21] Boadi D, Benchaar C, Chiquette J, Massé D. Mitigation strategies to reduce enteric methane emissions from dairy cows: Update review. Canadian Journal of Animal Science. 2004;**84**:319-335. DOI: 10.4141/A03-109

[22] Broucek J. Options to methane production abatement in ruminants: A review. Journal of Animal and Plant Sciences. 2018;**28**:348-364

[23] Nagaraja TG, Newbold CJ, Van Nevel CJ, Demeyer DI. Manipulation of ruminal fermentation. In: Hobson PJ, Stewart CS, editors. The Rumen Microbial Ecosystem. 2nd ed. London: Blackie Acad. Profess; 1997. pp. 523-632

[24] Mitsumori M, Sun W. Control of rumen microbial fermentation for mitigating methane emissions from the rumen. Asian-Australasian Journal of Animal Sciences. 2008;**21**. DOI: 10.5713/ajas.2008.r01

[25] McCaughey WP, Wittenberg K, Corrigan D. Methane production by steers on pasture. Canadian Journal of Animal Science. 1997;77:519-524. DOI: 10.4141/A96-137

[26] Diekert G, Wohlfarth G. Metabolism of homoacetogens. Antonie Van Leeuwenhoek. 1994;**66**:209-221. DOI: 10.1007/BF00871640

[27] Orskov ER, Ryle M. Energy Nutrition in Ruminants. London: Elsevier Applied Science; 1990. p. 149

[28] Thauer RK, Jungermann K, Decker K. Energy conservation in chemotrophic anaerobic bacteria. Bacteriological Reviews. 1977;**41**:100-180

[29] Lopez S, McIntosh FM, Wallace RJ, Newbold CJ. Effect of adding acetogenic bacteria on methane production by mixed rumen microorganisms. Animal Feed Science and Technology. 1999;**78**:1-9. DOI: 10.1016/S0377-8401(98)00273-9

[30] Benchaar C, Greathead H. Essential oils and opportunities to mitigate enteric methane emissions from ruminants. Animal Feed Science and Technology. 2011;**166-167**:338-355. DOI: 10.1016/j.anifeedsci.2011.04.024

[31] Dohme-Meier F, Machmueller A, Estermann L, Pfister B, Wasserfallen P, Kreuzer M. The role of the rumen ciliate protozoa for methane suppression caused by coconut oil. Letters in Applied Microbiology. 1999;**29**:187-192. DOI: 10.1046/j.1365-2672.1999.00614.x

[32] Machmuller A, Soliva CR, Kreuzer M. Methane-suppressing effect of myristic acid in sheep as affected by dietary calcium and forage

proportion. British Journal of Nutrition. 2003;**90**:529-540. DOI: 10.1079/BJN2003932

[33] Hristov AN, Varga G, Cassidy T, Long M, Heyler K, Karnati SKR, et al. Effect of *Saccharomyces cerevisiae* fermentation product on ruminal fermentation and nutrient utilization in dairy cows. Journal of Dairy Science. 2010;**93**:682-692. DOI: 10.3168/jds.2009-2379

[34] Newbold CJ, McIntosh FM, Wallace RJ. Changes in the microbial population of a rumen-simulating fermenter in response to yeast culture. Canadian Journal of Animal Science. 1998;**78**:241-244. DOI: 10.4141/A97-086

[35] Marden JP, Julien C, Monteils V, Auclair MR, Bayourthe C. How does live yeast differ from sodium bicarbonate to stabilize ruminal pH in high-yielding dairy cows? Journal of Dairy Science. 2008;**91**:3528-3535. DOI: 10.3168/jds.2007-0889

[36] McGinn SM, Beauchemin KA, Coates T, Colombatto D. Methane emissions from beef cattle: Effect of monensin, sunflower oil, enzymes, yeast, and fumaric acid. Journal of Animal Science. 2004;**82**:3346-3356. DOI: 10.2527/2004.82113346x

[37] Wright AD, Kennedy P, O'Neill C, Toovey A, Popovski S, Rea S, et al. Reducing methane emission in sheep by immunization against rumen methanogens. Vaccine. 2004;**22**: 3976-3985. DOI: 10.1016/j.vaccine.2004.03.053

[38] Murillo OM, Herrera TE, Corral LA, Pámanes CG. Effect of inclusion of graded level of water hyacinth on in vitro gas production kinetics and chemical composition of alfalfa hay based beef cattle diets. Indian Journal of Animal Research. 2018;(8):52, 1298-1303. DOI: 10.18805/ijar.11417

[39] Iñiguez L. The challenges of research and development small ruminant production in dry areas. Small Ruminants Research. 2011;**98**(1-3):12-20. DOI: 10.1016/j.smallrumres.2011.03.010

[40] Watanabe Y, Suzuki R, Koike S, Nagashima K, Mochizuki M, Forster RJ, et al. In vitro evaluation of cashew nut shell liquid as a methane-inhibiting and propionate-enhancing agent of ruminants. Journal of Dairy Science. 2010;**93**:5258-5267. DOI: 10.3168/jds.2009-2754

[41] Vasta V, Nudda A, Cannas A, Lanza M, Priolo A. Alternative resources and their effects on the quality of meat and milk from small ruminants. Animal Feed Science and Technology. 2008:223-246. DOI: 10.1016/j.anifeedsci.2007.09.020

[42] Chriyaa A. The use of shrubs in livestock feeding in low rainfall areas. Life support systems. Land Use, Land Cover and Soil Sciences. 2004:5

[43] Vandermeulen S, Ramírez-Restrepo CA, Beckers Y, Claessens H, Bindelle J. Agroforestry for ruminants: A review of trees and shrubs as fodder in silvopastoral temperate and tropical production systems. Animal Production Science. 2018;**58**:767-777. DOI: 10.1071/AN16434

[44] Tiemann T, Lascano C, Wettstein H, Mayer A, Kreuzer M, Hess H. Effect of the tropical tannin-rich shrub legumes *Calliandra calothyrsus* and *Flemingia macrophylla* on methane emission and nitrogen and energy balance in growing lambs. Animal. 2008;**2**(5):790-799. DOI: 10.1017/S1751731108001791

[45] Simone GO, Telma TTB, dos-Santos PM, Primavesi O, Frighetto R, Lima MA. Effect of tannin levels in sorghum silage and concentrate supplementation on apparent

digestibility and methane emission in beef cattle. Animal Feed Science and Technology. 2007;**135**(3-4):236-248. DOI: 10.1016/j.anifeedsci.2006.07.012

[46] Shreck A, Buckner DC, Erickson G, Klopfenstein T, Cecava JM. Digestibility of crop residues after chemical treatment and anaerobic storage. In: Nebraska Beef Cattle Reports MP. 2011. p. 94

[47] Ben-Salem H, Nefzaoui A, Makkar HPS. Towards better utilization of non-conventional feed sources by sheep and goats in some African and Asian countries. In: Ben-Salem H, Nefzaoui A, Morand-Fehr P, editors. Nutrition and Feeding Strategies of Sheep and Goats under Harsh Climates. Zaragoza: CIHEAM; 2004. pp. 177-187

[48] Salem H, Nefzaoui A. Feed blocks as alternative supplements for sheep and goats. Small Ruminant Research. 2003;**49**:275-288. DOI: 10.1016/S0921-4488(03)00144-5

[49] Mohammed ID, Baulube M, Adeyinka IA. Multi-nutrient blocks. I: Formulation and production under a semi-arid environment of north East Nigeria. Journal of Biological Sciences. 2007;**7**:389-392. DOI: 10.3923/jbs.2007.389.392

[50] Mohammadabadi T, Jolazadeh A. Replacement of alfalfa hay (*Medicago sativa* L.) with subabul (*Leucaena leucocephala*) leaf meal in diets of Najdi goats: Effect on digestion activity of rumen microorganisms. Tropical Animal Health and Production. 2017;**49**. DOI: 10.1007/s11250-017-1330-8

[51] Martine C, Copani G, Niderkorn V. Impacts of forage legumes on intake, digestion and methane emissions in ruminants. Legume Perspectives. 2016;**12**:24-25

[52] Yáñez-Ruiz DR, Martín-García AI, Weisbjerg MR, Hvelplund T, Molina-Alcaide E. A comparison of different legume seeds as protein supplement to optimise the use of low quality forages by ruminants. Archives of Animal Nutrition. 2009;**63**(1):39-55. DOI: 10.1080/17450390802611479

[53] Lüscher A, Mueller-Harvey I, Soussana JF, Rees RM, Peyraud JL. Potential of legume-based grassland-livestock systems in Europe: A review. Grass and Forage Science. 2014;**69**(2):206-228. DOI: 10.1111/gfs.12124

[54] Velez TM, Campos R, Sanchez-Guerrero H. Use of plant secondary metabolites to reduce ruminal methanogenesis. Tropical and Subtropical Agroecosystems. 2014;**17**:489-499

[55] Patra A, Saxena J. Exploitation of dietary tannins to improve rumen metabolism and ruminant nutrition. Journal of the Science of Food and Agriculture. 2011;**91**:24-37. DOI: 10.1002/jsfa.4152

[56] Naumann HD, Tedeschi LO, Zeller WE, Huntley NF. The role of condensed tannins in ruminant animal production: Advances, limitations and future directions. Revista Brasileira de Zootecnia. 2017;**46**(12):929-949. DOI: 10.1590/s1806-92902017001200009

[57] Becker PM, Van Wikselaar G, Franssen M, De Vos Ric M, Hall R, Beekwilder J. Evidence for a hydrogen-sink mechanism of (+)catechin-mediated emission reduction of the ruminant greenhouse gas methane. Metabolomics. 2014;**10**. DOI: 10.1007/s11306-013-0554-5

[58] Thakur M, Melzig M, Fuchs H, Weng A. Chemistry and pharmacology of saponins: Special focus on cytotoxic properties. In: Botanics: Targets and Therapy. 2011. p. 1. DOI: 10.2147/BTAT.S17261

[59] Patra A, Saxena J. The effect and mode of action of saponins on the microbial populations and fermentation in the rumen and ruminant production. Nutrition Research Reviews. 2009;**22**:204-219. DOI: 10.1017/S0954422409990163

[60] Crozier A, Jaganath IB, Clifford MN. Phenols, polyphenols and tannins: An overview. In: Crozier A, Clifford MN, Ashihara H, editors. Plant Secondary Metabolites Occurrence Structure and Role in the Human Diet. Chennai, India: Blackwell Publishing; 2006. pp. 1-24

[61] Yao LH, You-Ming J, Francisco SJT-B, Nivedita D, Riantong S, Shuang C. Flavonoids in food and their health benefits. Plant Foods for Human Nutrition. 2004;**59**:113-122. DOI: 10.1007/s11130-004-0049-7

[62] Sar C, Mwenya B, Pen B, Takaura K, Morikawa R, Tsujimoto A, et al. Effect on ruminal administration of *Escherichia coli* wild type or a genetically modified strain with enhanced high nitrite reductase activity on methane emission and nitrate toxicity in nitrate-infused sheep. British Journal of Nutrition. 2005;**94**:691-697. DOI: 10.1079/BJN20051517

[63] Sar C, Mwenya B, Pen B, Morikawa R, Takaura K, Kobayashi T, et al. Effect of nisin on ruminal methane production and nitrate/nitrite reduction in vitro. Australian Journal of Agricultural Research. 2006;**56**:803-810. DOI: 10.1071/AR04294

[64] Gamo Y, Mii M, Zhou XG, Sar C, Santoso B, Arai I, et al. Effects of lactic acid bacteria yeasts and galactooligosaccharide supplementation on in vitro rumen methane production. In: Takahashi J, Young BA, editors. Proceedings of the 1st International Conference on Greenhouse Gases and Animal Agriculture (GGAA), Obihiro, Japan, 7-11 Nov. 2001, Obihiro, Hokkaido, Japan. 2001. pp. 371-374. Journal of Animal Science. 2008;**86** (14 Suppl):E287-E292

[65] Saarela M, Mogensen G, Fonden R, Matto J, Mattila-Sandholm T. Probiotic bacteria: Safety, functional and technological properties. Journal of Biotechnology. 2000;**84**:197-215

[66] Sun-Waterhouse D, Peyta L, Wadhwa SS, Waterhouse G. Storage stability of phenolic-fortified avocado oil encapsulated using different polymer formulations and co-extrusion technology. Food and Bioprocess Technology. 2011;**5**. DOI: 10.1007/s11947-011-0591-x

[67] Tanvi S, Sun-Waterhouse D, Brooks J. Co-extrusion encapsulation of probiotic *Lactobacillus acidophilus* alone or together with apple skin polyphenols: An aqueous and value-added delivery system using alginate. Food and Bioprocess Technology. 2014;**7**. DOI: 10.1007/s11947-013-1129-1

[68] Harper JM. Food extruders and their applications. In: Mercier C, Linko P, Harper JM, editors. Extrusion Cooking. St. Paul, MN: American Association of Cereal Chemists, Inc; 1989. pp. 1-16

[69] Lele V, Zelvyte R, Monkeviciene I, Kantautaite J, Stankevicius R. Milk production and ruminal parameters of dairy cows fed diets containing *Lactobacillus sakei* KTU05-6 and *Pediococcus pentosaceus* BaltBio02. Polish Journal of Veterinary Sciences, Warsaw. 2019;**22**(2):327-335. DOI: 10.24425/pjvs.2019.129224

[70] Oliveira CA, Sousa DO, Penso JF, Menegucci PF, Silva LFP. Effect of different doses of a Bacillus-based probiotic on the in vitro digestibility of concentrates and forages. Journal of Animal Science. 2016;**94**(Suppl. 5):654. DOI: 10.2527/jam2016-1353

[71] Pinho RMA, Santos EM, de Oliveira JS, Gleidson GPC, Alves JP.

Relationship between forage neutral detergent fiber and non-fibrous carbohydrates on ruminal fermentation products and neutral detergent fiber digestibility in goats. Revista Colombiana de Ciencias Pecuarias Medellín. 2019;**32**(2):126-138. DOI: 10.17533/udea.rccp.v32n2a06

[72] Berhanu Y, Olav L, Nurfeta A, Angassa A, Aune J. Methane emissions from ruminant livestock in Ethiopia: Promising forage species to reduce CH_4 emissions. Agriculture. 2019;**9**:130. DOI: 10.3390/agriculture9060130

[73] Latham EA, Anderson RC, Pinchak WE, Nisbet DJ. Insight on alterations to the rumen ecosystem by nitrate and nitrocompounds. Frontiers in Microbiology. 2016. DOI: 10.3389/fmicb.2016.00228

[74] Storry EJ, Brumby EP, Tuckley B, Welch AV, Stead D, Fulford JR. Effect of feeding protected lipid to dairy cows in early lactation on the composition of blood lipoproteins and secretion of fatty acids in milk. The Journal of Agricultural Science. 1980;**94**:503-516. DOI: 10.1017/S0021859600028495

[75] Chengjian Y, Rooke J, Cabeza I, Wallace J. Nitrate and inhibition of ruminal methanogenesis: Microbial ecology, obstacles, and opportunities for lowering methane emissions from ruminant livestock. Frontiers in Microbiology. 2016;7. DOI: 10.3389/fmicb.2016.00132

[76] Vaikundamoorthy R, Zhen S, Hwang I. The potential role of secondary metabolites in modulating the flavor and taste of the meat. Food Research International. 2019;**122**. DOI: 10.1016/j.foodres.2019.04.007

Chapter 8

Water Use and Dairy Production System: An Indian Experience

G. Letha Devi, Anjumoni Mech, Sejian Veerasamy, Ravikiran Gorti and Mukund A. Kataktalware

Abstract

Increasing water scarcity and simultaneously growing demands for food and feed challenge agricultural production. Globally livestock feed sourcing is one of the major causes for water depletion; therefore, increasing livestock water use efficiency (LWUE) is necessary. There is a need to synthesise LWUE knowledge generated across different forage based livestock production systems (FLPS) over time and systematically identify entry points to enhance productive uses of freshwater resources. Although these systems vary by their degree of intensification, scale of water-related problems, and therefore in their values of LWUE, a number of common entry points to increase LWUE can be identified. To understand the pattern of livestock water use and social dynamics involved in water use and milk production, around 240 small and medium dairy farms in Karnataka, India, were used for the present study. Direct and indirect consumptive uses of water by animals considered were water used for drinking, water inputs through green and dry fodder, consumptive water usage for on-farm servicing and crop irrigation and water inputs through all upstream inputs such as medicines, vaccines and others. Water use efficiency (WUE) for production of milk alone is operationally defined in this study.

Keywords: water use efficiency, poverty, environment, livestock, socioeconomics

1. Introduction

Water is an essential component that is required in largest quantity by livestock. About 80% of animal water requirements is met by drinking water, and the rest of water needs are met through feed water. Production and reproduction performance of animals is directly affected by water availability and quality. Nonavailability of adequate water may cause adverse effects on animal growth and production. Water resources are shrinking day by day, and it warrants judicious use of water.

Milk production is challenged by increasing water scarcity and simultaneously growing demand for food and feed. Globally livestock feed sourcing is seen one of the major causes for water depletion, and therefore improvement in livestock water productivity is the need of the hour. Feed sources in smallholder production system largely consist of grazing, crop residue and concentrates, etc. Extensive smallholder systems in dryland ecoregions face the major challenge of water depletion for feed production. This demands better understanding of livestock-water interactions and designing strategies to improve water use efficiency (WUE).

Water use efficiency can be defined as the net return for a unit of water used. Improvement in water use efficiency aims at producing more food, income, better livelihoods and ecosystem services with less water. There is a considerable scope for improving water use efficiency of crop, livestock and other allied enterprises at field, thereby achieving sustainable food production. Water harvesting, supplemental irrigation, deficit irrigation, precision water application techniques and soil-water conservation practices are the bouquet of technology choices that we can resort to in achieving this goal. Practices not directly related to water management also impact water use efficiency because of interactive effects such as those derived from improvements in soil fertility, pest and disease control, crop selection or access to better markets.

However, we need to be cautious about achieving water use efficiency gains. Crop water use efficiency is quite high in highly productive regions, and yield (per unit of land area) does not necessarily correlate with water use efficiency in all cases. Water reuse within an irrigated area can compensate for the perceived losses at the field in terms of water quantity, but that will not be of any help in maintaining the water quality. We need to create an enabling environment for enhancing water use efficiency by farmers in field. Apart from this, we need a thorough understanding of the biophysical environment as well as social and economic dynamics existing between different elements of farm and field.

While identifying priority areas for bringing in improvements in water use efficiency and formulating strategies and action points for bringing in substantial improvements in water use efficiency, the following points have to be considered: (i) high-poverty less water efficient areas, (ii) water-scarce areas, (iii) areas neglected for development of water resources, and (iv) areas of faster water resource depletion. However, these are huge challenges to be achieved, and strategies need to be evolved keeping in view complex biophysical, social and economic factors.

2. Water footprint

Water footprint is defined as the extent of water use in relation to consumption of goods and services by people. In a broader sense, a country's water footprint is the volume of water required for the production of the goods and services used for direct and indirect consumption by the population of the country. Water footprint can be of two types: (i) internal water footprint or water used from internal or domestic resources and (ii) external water footprint or water used to produce imported goods and services. The USA has an average water footprint of 2480 m^3/cap/year, and China has an average footprint of 700 m^3/cap/year. Global average water footprint is 1240 m^3/cap/year. Any country's water footprint is determined by factors such as consumption volume (with respect to gross national income); consumption pattern; climate; and water use efficiency of agriculture and allied sectors.

The water footprint gives an account of amount of water used to produce each of the goods and services we use. It can be measured for a single process, such as growing a crop, for a product and fuel, etc. It also gives an idea about volume of water being consumed by a country in a specific basin or from a specific source. The water footprint looks at both direct and indirect water use of a product. It includes water consumption and pollution throughout production cycle from supply chain to consumer.

Water footprint can be measured in terms of per unit of goods produced and per hectare of area under crops or in any other functional units. This also gives us an idea about different uses of our limited freshwater resources and the ways and means by which they get polluted. If the water is sourced from a water-scarce area, the impact of low water productivity to high water footprint can be significant and require immediate attention.

For the purpose of quantifying its use, water can be divided into three components: green, blue and grey. The three components together provide a comprehensive picture of water use by demarcating water source, either as rainfall, groundwater, or surface water, apart from freshwater requirement for removal of pollutants, to make it reusable.

3. Types of water footprint

1. **Green water footprint**: water from precipitation/rainfall that is accumulated in deep soil and includes the evapotranspiration component and water incorporated by plants. This is the most relevant water component for agricultural and allied products.

2. **Blue water footprint:** surface water or groundwater resources and is either evaporated or incorporated into a product across a temporal and spatial regime. Irrigated cropping, industry and domestic consumption of water falls into blue water footprint.

3. **Grey water footprint:** volume of freshwater essentially needed to remove pollutants and make it reusable. This component takes into account point source pollutants discharged to any freshwater source directly or indirectly or other diffuse sources.

Livestock plays a vital role in supporting rural livelihoods in the Indian context. At the same time, there are growing concerns regarding highly water-intensive operations in livestock rearing, which is considered as one of the major enterprises for water depletion and putting huge pressure on depleting and water-scarce resources. In forage-based livestock production systems, be it grazing, mixed-irrigated or mixed-rain fed, feed sourcing is largely contributed from pasture or crop residue. In dryland areas of arid and semi-arid ecosystems, extensive forage-based livestock production systems are in place, and in such situations, water used for feed production is a major concern. Thus, such situations warrant the pressing need for understanding the livestock water dynamics and better strategies and framework for developing comprehensive entry points to improve livestock water use efficiency.

Based on global experiences from different livestock production systems, the entry points for improving livestock water use efficiency can be categorised into different groups, based on their operational limits. They are:

i. Feed water productivity.

ii. Feed sourcing and feeding management.

iii. Livestock feed use efficiencies.

iv. Institutions to create enabling environment, for better water use management.

v. Market linkages for bringing out water saving technologies to consumers.

4. LWUE in forage-based livestock systems: challenges and opportunities

In the major **forage-based livestock systems** like grazing, mixed-rainfed and mixed-irrigated systems of dryland production environments, the basic objectives of production as well as intensity of production operations have a great diversity within and among those systems [1]. This diversity creates many challenges for water efficiency of these livestock production systems. This creates implications and prospects at the same time for achieving efficient water use in such production systems. To elaborate further, dry and green fodder constitute major feed component in dryland production systems. Like in the case of the most intensive systems, say mixed-irrigated production system as practised in India, concentrate feed use does not exceed 10% [2]. Feed acts as a major interface between water and livestock, and such diversity in managing feed sourcing and feeding practices poses challenges and implications for the type, scale of importance and method of quantifying and strategising livestock water use efficiency.

Strategies to improve quality of locally available feed and feed management are core to any framework to improve livestock water use efficiency in any production system. We need to focus on activities like selection of crops, intercropping for maximum land and water utilisation, urea treatment of crop residues, chopping of coarse residues, etc. In mixed-irrigated systems, an improvement of feed quality (from 7 to 8.5 ME MJ kg^{-1}) can lead to saving of >50 m^{-3} of water/cow/year [2]. Similarly, in mixed-rainfed systems, urea treatment of crop residues led to a considerable improvement in livestock water use efficiency [3–5]. While considering better animal management practices, livestock water productivity (LWP) can be enhanced, by reducing animal's energy requirement by means of limiting animal movement, especially in peak summer seasons. Descheemaeker et al. [3] reported that in mixed-rainfed systems, approximately 12% of the metabolisable energy of animals is spent for walking long distances for feed and water. This energy loss can be avoided by better feed sourcing and feed management.

5. Method of assessment of livestock water use

An effort was made to assess and analyse LWUE in smallholder and commercial production and to formulate for strategies for improving LWUE. Primary data was collected from small- and medium-sized dairy farms in Kolar and Shimoga district, Karnataka, India. The total sample size was 240 dairy farms. The consumptive use of blue water (direct and indirect) was assessed using primary data through personal interview and observation in particular farms. Primary data from smallholders and commercial dairy units in Kolar and Shimoga district of Karnataka, India, were collected. Water use efficiency (kg/animal) was estimated and compared for smallholder as well as commercial dairy production systems using the following formula:

$$WUE = (Y/U)*100 \tag{1}$$

where Y = Marketable yield (kg/animal) and U = Seasonal consumptive use of water (m^3).

Water use efficiency for crop biomass used as fodder = Total Biomass/water applied at different levels of requirement [6] method was used for calculation of LWP of feed (recommended by the IWMI). Different water wastage points in different operations were identified, and strategies to reduce water wastage were formulated using participatory focus group discussions.

The major challenges associated with LWU as perceived by farmers were analysed and ranked based on rank coefficients. Scarcity of water for livestock drinking, other livestock operations and feed quality due to low water quality used for crop production were the major challenges across all the seasons (**Tables 1** and **2**).

The water intake by animals through forage and other feed ingredients is more as compared to water intake through drinking water and that used for on-farm servicing operations such as cleaning, etc. The average direct consumptive water use by smallholder system was found to be 97 litres per day and 127 litres per day for commercial dairies. The calculated water use efficiency for smallholder system was 0.85, and for commercial dairying it was 1.62. The water use efficiency was more in the case of commercial dairy farming and less in the case of smallholder production system.

There are various factors affecting water use by livestock. The major factors are seasons, different weather parameters, fodder, feed and other inputs. The source of

Key LWU-related problems	Seasonal variations								
	Summer			Winter			Rainy		
	1	2	3	1	2	3	1	2	3
Scarcity of water for livestock drinking	✓				✓				✓
Scarcity of water for livestock operations	✓				✓				✓
Scarcity of water for feed production					✓				✓
Inefficient use of available water			✓			✓	✓		
Soil/nutrient loss	✓					✓		✓	
Poor feed/fodder quality	✓					✓			✓
High feed scarcity	✓				✓				✓
Use of common property resources	✓			✓			✓		
Postharvest feed quality and quantity	✓				✓				✓

Table 1.
Problem matrix showing the scale of importance of LWU-related problems across seasons.

Operations	Smallholder system	Commercial dairying
Drinking	40	52
Washing shed	55	90
Washing animals	25	38
Cleaning cans and other equipment	10	25
Water contained in feed and fodder	743	740
Total	**873**	**945**
Milk yield/day/animal	7.4	15.4
WUE = (Y/U)*100	**0.85**	**1.62**

Table 2.
Direct and indirect water use (litre/day/animal/kg of milk) and WUE in different dairy production systems (n1 = 200, n2 = 40).

Factors	Rank
Seasonal variation	I
Weather parameters (temperature, rainfall, humidity)	II
Fodder, feed and other inputs	III
Source of water (bore well, canals, ponds, etc.)	IV
Animal conditions	V
Animal output	VI

Table 3.
Factors affecting water use (ranking; n = 240).

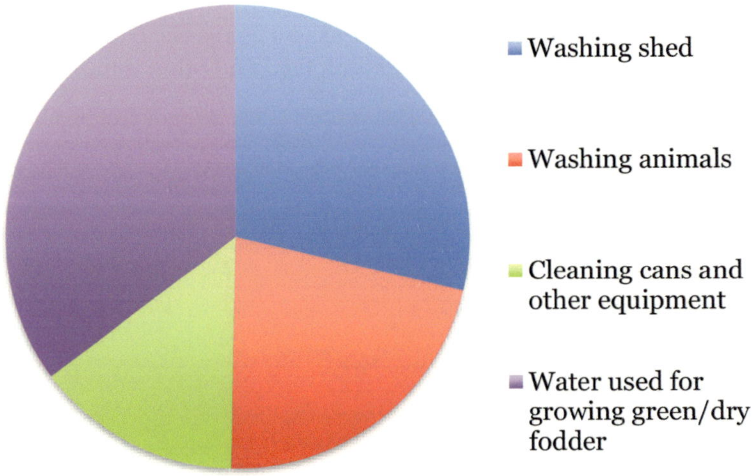

Figure 1.
Perceived water wastage points in summer season (%respondents), n = 240.

water and animal conditions like lactation stage, age and body and health conditions also play a role in water use efficiency (**Table 3**).

The water wastage points mainly in summer season were identified, which is presented in **Figure 1**.

6. Conclusion

Water availability and quality are the major challenges that are faced by the livestock and crop production systems in recent times. The observations in the study show that water inputs through forage and other feed ingredients are more than the water inputs through drinking water and that used for on-farm servicing operations such as cleaning, washing, etc. Proper management strategies are highly essential for sustaining the livestock production systems and meet the food demands of a growing population with the available water resources, for which water saving technologies and strategies are the need of the hour.

Author details

G. Letha Devi[1]*, Anjumoni Mech[1], Sejian Veerasamy[1], Ravikiran Gorti[1]
and Mukund A. Kataktalware[2]

1 ICAR-National Institute of Animal Nutrition and Physiology, Bangalore, India

2 ICAR-National Dairy Research Institute, Bangalore, India

*Address all correspondence to: lethaayur@gmail.com

IntechOpen

© 2020 The Author(s). Licensee IntechOpen. This chapter is distributed under the terms of the Creative Commons Attribution License (http://creativecommons.org/licenses/by/3.0), which permits unrestricted use, distribution, and reproduction in any medium, provided the original work is properly cited.

References

[1] Seré C, Steinfeld H. World Livestock Production Systems: Current Status, Issues and Trends. Rome: FAO Animal Production and Health Paper 127; 1996

[2] Haileslassie A, Blummel M, Clement F, Ishaq S, Khan MA. Adapting livestock water productivity to climate change. International Journal of Climate Change Strategies and Management. 2011;**3**:156-169

[3] Descheemaeker K, Bossio D, Amede T, Ayalneh W, Haileslassie A, Mapedza E. Analysis of gaps and possible interventions for improving water productivity in crop-livestock systems of Ethiopia. Experimental Agriculture. 2011;**47**:21-38

[4] Haileslassie A, Blümmel M, Murthy MVR, Samad M, Clement F, Anandan S, et al. Assessment of livestock feed and water nexus across mixed crop livestock system's intensification gradient: An example from the indo-Ganaga Basin. Experimental Agriculture. 2011;**47**:113-132

[5] Peden D, Tadesse G, Misra A. Water and livestock for human development. In: Water for Food, Water for Life: A Comprehensive Assessment of Water Management in Agriculture. Earthscan, Cambridge Publishing, Oxford University Press; 2007. pp. 485-514

[6] Mekonnen MM, Hoekstra AY. The green, blue and grey water footprint of crops and derived crop products. Hydrology and Earth System Sciences. 2011;**15**:1577-1600